APOLLO'S FIRE

*Darwin's Orchestra: An Almanac of Nature
in History and the Arts*

*Adam's Navel: A Natural and Cultural
History of the Human Form*

AS EDITOR

The Annotated Archy and Mehitabel

Arsène Lupin, Gentleman-Thief

APOLLO'S FIRE

{ *A Day on Earth* }
{ *in Nature and Imagination* }

MICHAEL SIMS

VIKING

VIKING

Published by the Penguin Group

Penguin Group (USA) Inc., 375 Hudson Street, New York, New York 10014, U.S.A. · Penguin Group
(Canada), 90 Eglinton Avenue East, Suite 700, Toronto, Ontario, Canada M4P 2Y3 (a division of Pear-
son Penguin Canada Inc.) · Penguin Books Ltd, 80 Strand, London WC2R oRL, England · Penguin
Ireland, 25 St. Stephen's Green, Dublin 2, Ireland (a division of Penguin Books Ltd) · Penguin Books
Australia Ltd, 250 Camberwell Road, Camberwell, Victoria 3124, Australia (a division of Pearson Aus-
tralia Group Pty Ltd) · Penguin Books India Pvt Ltd, 11 Community Centre, Panchsheel Park, New
Delhi – 110 017, India · Penguin Group (NZ), 67 Apollo Drive, Rosedale, North Shore 0745, Auck-
land, New Zealand (a division of Pearson New Zealand Ltd) · Penguin Books (South Africa) (Pty)
Ltd, 24 Sturdee Avenue, Rosebank, Johannesburg 2196, South Africa

Penguin Books Ltd, Registered Offices: 80 Strand, London WC2R oRL, England

First published in 2007 by Viking Penguin, a member of Penguin Group (USA) Inc.

10   9   8   7   6   5   4   3   2   1

Copyright © Michael Sims, 2007
All rights reserved

LIBRARY OF CONGRESS CATALOGING-IN-PUBLICATION DATA
Sims, Michael,———
    Apollo's fire : a day on Earth in nature and imagination / Michael Sims.
        p. cm.
    Includes bibliographical references and index.
    ISBN 978-0-670-06328-4
    1. Time—History. 2. Days. 3. Astronomy—Miscellanea. 4. Astrology and mythology.
    5. Astronomy—Observations. I. Title.
    QB209.S56 2007
    529'.1—dc22   2007006024

Printed in the United States of America
Designed by Carla Bolte · Set in Celeste

To Laura, of course,

with love and admiration

The first in time and the first in importance of the influences upon the mind is that of nature. Every day, the sun; and after sunset, Night and her stars.

—Ralph Waldo Emerson

The moon and sun are eternal travelers. . . . Every day is a journey, and the journey itself is home.

—Matsuo Bashō

The creation of the world did not occur at the beginning of time; it occurs every day.

—Marcel Proust

CONTENTS

Evening

Night

Shadow Puppets on the Moon

I was in the fifth grade when our teacher demonstrated the relationship between the sun and the turning earth. He switched off the overhead lights and lowered the creaking windowshades, making the classroom dark enough to enable him to prove his point. He placed a school globe on his desk and set a flashlight on a nearby table. When he clicked on the flashlight, its beam illuminated the local atmosphere: swirling motes of chalk dust. Thirty young students sat in the theatrical darkness, watching their spotlit planet like gods. With one finger the teacher spun the globe and it rotated slowly on its tilted axis.

As far back as I could remember, I had pored over maps and atlases. They had revealed that the creeks and hills of rural Tennessee, where I was born, were only the local miniature version of a larger pattern of wrinkles on the surface of the earth. Encyclopedias had shown me the changing borders—how our beloved imaginary lines conga back and forth across the patient realities of mountain and coastline. From Mercator's flat distortions on maps I had tried to envision my rotund home planet. Once, with a ballpoint pen, I drew a cartoon outline of continents on an orange's resilient pored surface and then peeled the fruit and tried to spread out the peel on the table. Its curves had to be flattened, and doing so broke up the lines of the continents. Aha: so this is how it works.

But until this moment in the dark classroom I had never felt the dizzy thrill of walking around on a wet boulder rolling through the black night of space. How could so much water cling to the surface of a ball? I felt a primitive shiver of awe as I crowded close to the globe to peer at the line of demarcation between sunlight and shade. I saw Thailand emerging into light. Then the sun was coming up over the Bay of Bengal, warming the coast of Sri Lanka, waking the sleeping people of Nagpur, stirring a dawn breeze in Karachi. I imagined elephants rising from slumber with elderly stretching fore and aft. I waited; the globe spun lazily. It seemed a long time before my own part of the world emerged from night and into day. I leaned closer. My town was invisible; my county, even my state were not large enough to deserve a name from this spaceship point of view. Yet I was down there somewhere, the size of an atom, waking up in my bed with the sun's light coming through the curtains as Earth turned toward it. Every morning the elephants woke up and so did I. Every morning the sun returned.

☀

Already I had been interested in astronomy for some time, in a lazy, casual way, and I had read about this demonstration of how to visualize the interaction between the sun and the

Earth. What I did not foresee was that, besides always being precisely half in shade itself, besides showing the predictable daylight and nighttime hemispheres, the sidelit globe also cast a blurry round shadow against the far wall of the classroom. I remember thinking, *Earth has a shadow!* I began to understand eclipses for the first time, saw Earth's umbra fading into empty space as the planet revolved in its orbit, until now and then it chanced to fall across the moon's face. In time I learned that our planet's round shadow eclipsing its satellite was one of the earliest bits of evidence for a globular instead of a flat Earth. Years later I stood among skygazers in a Nashville field, awaiting a total eclipse of the moon. Finally we saw the shadow of the planet begin to creep across the familiar old face in the sky. But the sun hadn't been down for long. I felt that I ought to be able to hold up my hands and cast shadow puppets on the moon: Look, a rabbit! A dog!

Almost four decades after the classroom demonstration, my experience of our turning globe is more personal, less theoretical. I have flown many times at the borderline of the atmosphere and outer space, with curving darkness above and the curving wrinkled Earth below. Nowadays— especially on barefoot summer days—I feel the planet turning beneath my feet when I walk. I balance atop it like a circus acrobat. And every place where I have watched the sun rise and set, where I have admired the moon's slow climb or looked out a bedroom window at the stars over my bed, I have yearned to write a book that would capture some of my own response to this majestic daily rhythm.

✳

Another childhood memory comes to mind when I consider these themes. I recall a winter nighttime in the country, still and quiet, with snow re-imagining the normally dark fields and yards with its fanciful illumination. I was bundled for the Arctic. My jacket's hood was cinched tight around my face, barely leaving my eyes and nose and mouth exposed. A moment before, I had been indoors, helping decorate the Christmas tree and watching holiday cartoons on TV—flying reindeer, singing snowmen. It had been a cozy scene, familiar and comprehensible. But I had been unable to resist the new look of the night glimpsed through a finger-melted peephole in the frosted windowpane. Now I sighed clouds as I crunched through the snow and peered at this alien world like an exploring astronaut.

I admired the tracery of branches and twigs precisely inked on the snow. Then I walked out from under the trees and stood facing a white duvet that had transformed our low stone wall into a comfortable-looking pillow, the whole expanse virgin except for a scribbled trail of rabbit prints near my boots. The white landscape seemed impossibly bright, and countless tiny sparkles winked at me as I moved. I stopped and peered more closely. Why this diamond glitter? I knelt down and reached out a gloved hand to touch the snow. The shadow of my arm that accompanied this gesture was so black and sharp that it surprised me, and involuntarily I looked up.

Never before had I seen so many stars. It was their light that had made the snow so luminous and that winked in its crystals. Winter's air is the most pure, cleansed of autumn's dust and pollen. Today I know that I could have observed no more than a couple of thousand stars, but at the time it seemed like millions shining above me—different luminosities, different colors, singly and in patterns, and behind them the pale river of the Milky Way. I stared until my neck complained. Then I sat on the squeaky snow and finally lay on my back and faced the sky as if it were the ceiling of my bedroom. I was so bundled the cold couldn't threaten me. My breath hovered over my face and I peered upward through it.

I had already discovered the cosmic vistas of science fiction. I knew what I was seeing. I was seeing how the universe looks from a spaceship. Then I realized what I ought to have learned in the classroom demonstration with flashlight and globe: that the Earth *is* a spaceship, wrapped in a fragile atmosphere, hurtling through the very definition of the void. The longer I stared upward, the more I lost my surroundings, the weight of my own body, my connection with the planet where I was born. I lost the memory of whatever small rituals I had been performing a few minutes ago in my inconsequential life. The stars seemed to move. There was no boundary to them, no safe frame to keep them distant. I fell into the sky.

The sensation of vertigo terrified me. I called out, or tried to, but instead of sitting up I scrambled at the snow with my outstretched arms. I had to cling to something to keep from falling, and of course I grasped at the earth curving beneath me. Finally I was able to close my eyes and shut out the beckoning sky. I didn't open them again until I was sitting upright on the snow, and as I rose to my knees and then my feet I carefully avoided looking upward. I peered toward our house—but it wasn't there. For a moment there was no cloud of breath before my face: *I'm lost!* But I was simply facing in the wrong direction, and when I turned around I saw the friendly yellow squares of light no more than a hundred feet away. Raising my eyes to look for the house, however, had brought the lower stars into my range of vision. Only those above the awning of my eyebrows were still invisible. I held my hand up and shaded my eyes as if guarding them from the sun. Then, looking down at the safe ground, and carefully not thinking about the many distant suns whose nuclear fire had traversed space since the days of the dinosaurs to sparkle in the snow crystals, I ran as hard as I could toward the small familiar world of my home.

✻

Perhaps these experiences contribute to the sense of affection with which I watch the moon's progress throughout the night, or the quiet satisfaction I find in watching shadows shrink and grow again as the sun arcs over from morning to evening. At the end of the day I sleep best after I have walked or at least stepped outdoors and looked up at the sky. Even an upward glance out the window reassures me that the calamities and scurry of normal urban life in the new millennium still take place under the same vault of stars that watched the painters emerge from the cave at Lascaux. The moon outside the train window illuminated Sappho's heartbreak; the sun glaring on the windows of new office buildings cast Alexander Pope's bent shadow ahead of him. We find our context where we can. Mine resides in the daily rhythms of my planet and its star.

This book, however, is not about my personal experiences. After these few minutes at the microphone, I will vanish backstage. I find a distillation of dawns and sunsets more relevant to my theme than a diarist's account of a particular day in a particular location. This method will also permit me to call upon people from around the world and throughout history for their experiences and insights. I will draw from both natural and cultural history, and I can't always tell where one ends and the other begins. Henry David Thoreau wrote in his journal that he kept two notebooks, one for facts and one for poetry, but that he found it difficult to preserve the distinction between these artificial categories, "for the most interesting and beautiful facts are so much the more poetry and that is their success." More concisely, Samuel Taylor Coleridge said that he attended the Royal Institution lectures on science primarily to renew his stock of metaphors.

Still, this is a personal and subjective book. I intend to explore the day from beginning to end, creating what I hope will be an informative and entertaining companion on our shared journey. Although we will travel from dawn to nighttime, the story describes a distilled, ideal day, rather than a particular one. Clouds appear overhead when I need them and then obediently depart, and a full moon vanishes from one chapter to the next to make room for the stars. In *Apollo's Fire* I have chosen my topics with the same goal that Dickens said motivated him in writing *Bleak House*: "I have purposely dwelt upon the romantic side of familiar things." Romance, like beauty, is in the eye of the beholder.

Morning Twilight

Gradually the dark sky paled until it looked like far away smoke. There was no colour anywhere.... It felt more like dusk than dawn, but not really like any time of day or night.

—Dodie Smith, *I Capture the Castle*

Stephen Dedalus's Cosmic Address

Early in James Joyce's *Portrait of the Artist as a Young Man*, the youthful protagonist sits distractedly in a classroom, writing in the flyleaf of his geography text. His consciousness moves ever outward until he assigns himself a kind of cosmic address:

> *Stephen Dedalus*
> *Class of Elements*
> *Clongowes Wood College*
> *Sallins*
> *County Kildare*
> *Ireland*
> *Europe*
> *The World*
> *The Universe*

Stephen was not the first student to indulge in such exercises. They impart a helpful perspective on our place in the scheme of things—especially when you are a child trapped in a classroom and the whole grand world beckons alluringly from outside the window. Too often organized education packages information in units that aid memorization and regurgitation. Such facts are useful, but because the brain thinks symbolically and narratively we also need examples that demonstrate the information's relevance to our daily lives.

Like his creator, Stephen Dedalus gropes for connections that clarify meaning and significance. Whatever else he may have been missing while his mind wandered, young Stephen was at least thinking about one of the essential questions of human existence: our place in the universe. A geology text is a fitting site for such musings. Many questions about the meaning of life grapple with the idea of context. Where are we in relation to the rest of the cosmos, to other sentient creatures, even to our fellow human beings? Stephen's own list locates him at the heart of concentric circles moving outward from the individual to the entire universe.

"The child transfigured the blowsy lawns with his playing," writes Dylan Thomas in one of his stories. We all transfigure the world with our playing. *Homo sapiens* has been playing for longer than we can envision. We transfigure our surroundings, our individual and collective experience, with playing and knowledge and ignorance. Ptolemy and Copernicus were both correct: our universe—

our physical and mental universe—is centered around both the Earth and the sun.

The cycle of the rising and setting sun, the alternation of light and dark, is the oldest story on Earth. It may well be the genesis of all linear narrative. Before our ancestors tamed the fire that lightning brought down from heaven, before they planted seeds and lingered in one place long enough to watch them magically grow into crops, before they even understood the connection between sexual intercourse and childbirth, they told the story of the day itself. This primordial cycle still determines the rhythms of our lives, our jobs, our holidays, our weather.

Mythology and science both suggest that, as far as we are concerned, the real purpose of the day is to show us our place in the cosmos. The day is an astronomical phenomenon and a unit of time, and as a frequently repeated experience it became a symbol long ago. But it is also one of our favorite ways of reinserting ourselves into a larger narrative. Every culture has conjured myths about the origin of the day and night. The sun climbing above the horizon is one of the earliest pictographic symbols in ancient history, and it reappears in most children's first drawings. It is a milestone in each family when a child begins to understand the concept of tomorrow morning and last night. What a reassuring idea to a child that even if darkness demands a nightly surrender to monster-haunted unconsciousness, a shiny new day will greet her upon awakening. The day is a small enough unit of time to be easily comprehended by both societies and individuals. Gathered into two other natural cycles, lunar months and solar years, it forms the basis of all calendars. The story of how we learned what is actually happening in this pattern is the story of science coming of age. And, because imagination never waits until the facts are in, every culture is rich in myths about this most easily observed and most frequently occurring natural cycle on Earth—the sun's dependable journey across the sky from dawn to dusk.

"We are nature seeing nature," remarks the American writer Susan Griffin. "We are nature with a concept of nature." Apparently the chief difference between *Homo sapiens* and our manifold kin is that for us nature is symbolic. Our mental picture of the world is more than the map of surroundings hanging over the head of an animal—a scent-based map in dogs, perhaps, and sound-based in bats. No doubt blue whales and Arctic foxes and birds of paradise all have their own concept of nature, their mental picture of the world and the interrelationships of its many parts, but presumably they do not invent for themselves a story about it as they journey through their days.

Indeed, the very word *journey* preserves the importance of this daily rhythm. Digging back through Middle English and Old French, we find it deriving from the Latin *diurnata,* for "a day's work." Over the centuries many cultures described distances as the number of days of walking or riding necessary to reach a destination, just as nowadays gadabouts on a shrinking planet refer to distances in terms of hours of driving or flying. The Latin *dies,* "day," begat *diurnus,* "daily," and the Late Latin *diurnalis.* From the last word we get *journal,* which as surely as *diary* refers to a record that is kept on a daily basis. Hidden treasure rewards even the most casual peek into such seeming trivia. For example, the beautiful word *ephemeral,* which we use to mean transitory or evanescent, actually comes from two Greek words meaning "lasting but a day." Our daily routines and chores, our waking and sleeping hours, are literally ephemeral even though they relive the experience of the ages.

Darkness

To fully appreciate the arrival of daylight, we need to imagine the darkness that, until recent centuries, only dawn could banish. No one in the industrialized world can remember so far back, but let's try to envision it. Even recollections of wilderness camping or a nighttime power failure will aid this mental effort. Tonight we will look more closely at our experience of darkness, but for now let us try to conjure what life was like for human beings before the flick of a switch illuminated a room, before sentry lights guarded homes, before headlights ran ahead of vehicles like torch-bearing heralds. Go back before your great-grandparents electrified their homes. Picture a time when no artificial lights erased the lesser stars of every constellation and no glow of cities could be seen from space by astronauts circling the Earth. Rewind the film of history and watch nighttime lights winking out around the globe.

There was a time when Jack the Ripper lurked in the flickering shadows of gaslit London streets, when frightened citizens thought twice before plunging into the darkness between two oases of light. A century earlier a group of Englishmen—including Josiah Wedgwood, founder of the pottery dynasty, and Erasmus Darwin, grandfather of Charles—called themselves the Lunar Society, because to provide illumination for their tipsy progress homeward they timed their meetings to coincide with the full moon. The arrival date of Easter varies each year for the same reason: because pilgrims needed light to travel by. Ever since the Council of Nicaea in the fourth century, Easter occurs on the first Sunday after the full moon following the vernal equinox. But the moon is unreliable illumination; it wanes, hides behind clouds, even disappears for a few nights each month. Its borrowed light is feeble compared with the sun's.

For centuries in Europe, the time between midnight and cock-crow was considered "dead time," and apparently this period inspired the phrase "dead of night." According to Horatio, the watchmen saw the ghost of Hamlet's father during "the dead vast and middle of the night." It was said to be the likeliest time for an attack by demons, rogues, or werewolves. The last few hours of darkness before daylight really do seem the slowest. During a night of pain, the aggrieved and the lonely and the desperately ill wonder if they can survive until dawn. In ancient rituals of symbolic death that led to rebirth, darkness often represented not annihilation but latent being, the period of waiting before the light of birth.

But we heirs of linear rather than cyclical myths imagine death as a final darkness. For us, darkness is the time before creation, the godless time. Go back before we banished it.

Imagine the eons of nightly obscurity before our protohuman ancestors dared to carry back to the cave a tree branch that had been ignited by fire from heaven. Think of the conceptual landmark that our ancestors reached when terror of fire was replaced by the idea of taming it.

Darkness is upon the face of the Earth. It is dark now and it will remain dark until morning and there is nothing that we can do about it. The darkness seems unending and impenetrable. We know from bloody experience that it is full of predators who find our flesh tasty, beasts that can see in the dark and smell our fear. This danger is not a projection of our imagination into the mysterious darkness. The danger is real. Can you imagine that you and your family are huddled together in a cave, reassuringly holding hands in the darkness? Are you looking up anxiously at every rustle of leaves?

Only if we can imagine their fear can we begin to understand how grateful our ancestors were for the dawn.

The Threat of Dawn

In 1923, in the early morning of his own life and that of the twentieth century, the already brilliant Argentine writer Jorge Luis Borges wrote a poem entitled "Amanecer" (Break of Day). Just before sunrise, "daunted by the threat of dawn," he recalls the frightful conjecture of such philosophers as Arthur Schopenhauer and Bishop Berkeley—that the world doesn't really exist, that all we imagine we see and hear around us is a dream, a projection, a common delusion. (To which philosophical dead end Samuel Johnson famously replied by kicking a stone and snarling to Boswell, "Thus I refute Berkeley!") As the sky lightens and the sun finally arrives, Borges speculates that, if the teeming streets of Buenos Aires are no more solid than a dream, there is a moment when their airy existence is in grave danger, "the shuddering instant of daybreak." This is the time of day when fewer and fewer people are left asleep to keep dreaming the world. Only those who remember its outlines from yesterday can be assured that they are facing the same reality. As the sun comes up, Borges begs his house to continue existing, and somewhere out in the shared delusion a bird begins to sing. But the now banished night, writes Borges, lingers on "in the eyes of the blind." He had not yet succumbed to the increasing difficulties and eventual blindness that would darken his later years. Eventually the sun stopped rising for him and only the night remained.

A Light in the East

More than 70 percent of our planet is covered in water, and hundreds of millions of people around the world experience the arrival of each day along the boundary between earth and sea. So let us begin our experience of the day at a shoreline.

On the eastern coast of the United States, light creeps across the Atlantic from Africa and Europe. The sun wakes pelicans and jellyfish before wading ashore to warm the footprinted sand. In the few coastal areas where an early riser can escape some percentage of the omnipresent light pollution, clouds may not reflect enough illumination from the surface to reveal their presence. Dark clouds stealing across an equally dark sky generate a new mythology as they erase part of Orion or the Scorpion. In this situation, arriving well ahead of color, dawn begins as the slow resolution of the sky into dark masses of shadow against the gradually lightening heavens behind them.

Now it is light enough that cloud shapes are visible all across the sky, even in the west, above the curving earth where people and animals still slumber. Often at the seashore clouds can be so wind-tossed that their shadowy dawn shapes fade out at the edges into a charcoal sketchiness. As they float grandly toward you and then over your head, they look like torn shreds of dark cotton. The military procession of this army of giants would feel ominous at dusk. But dawn seldom feels ominous; we have been too long conditioned. There is no knowledge more closely woven into the fractaling neurological pathways of the brain than the knowledge that dawn will be here soon to change everything about the world. The boundaries of our daytime world are being restored—intact, freshened.

※

Dawn cloud shows are some of the most dramatic of the day. The lowest level of clouds, those just coming over the horizon, remain dark, finally purpling along their top edges. Swaths—areas still too vague to be called "bands"—are salmon-colored, and farther up rise washes of white and pink and purple. Above them blue is quickly pouring into the formerly black sky, diluting from indigo through cobalt and cerulean. The dark forms of the clouds—constantly changing shape over a coastal area where differing degrees of warmth keep air moving—merge and twist, expand like amoebae, and chase each other overhead.

The increasing red hues in the east, fading to the lighter end of the impressive range of celestial blues, remind us that as the planet turns your particular neighborhood toward the sun, light rays are angling through a great mass of atmosphere. It is easy to forget that sunrise is not a solar phenomenon but instead purely terrestrial. Our local star interacts no more with the parade of excitable clouds heralding its arrival than do the headlights of a distant car with the rain that blurs them on your windshield.

The atmosphere's effect on light rays will be a topic that recurs throughout this book, even after dark, so we ought to begin at the beginning. People often ask why sunrise and sunset colors are so dramatic, but we forget to ask why the midday sun seems yellow to us although astronauts and space cameras report it as white. The atmosphere cushions us too well to permit light rays to sneak through without alteration; there is a great deal of invisible substance floating around up there. Even without the larger particles suspended in it, air consists of molecules of oxygen, nitrogen, hydrogen, and other elements. The wavelength of light is at least a thousand times larger than any individual molecule, however, so their tiny atoms alone are not the primary obstacle to incoming photons. (*Wavelength* refers to the distance between crests as the wave moves through space or through some medium such as air.)

The larger culprit is the air's dense population of dust and pollen and floating microorganisms. To visualize how strong this mass can be, remember that air blurs distant views like a screened window, that it buoys up jets and albatrosses, that it is capable of incinerating a meteor through friction or of stirring enough speed to toss a yacht like a bathtub toy. Textbooks often compare our planet's atmosphere to the peel of an orange, to give us some idea how relatively thin it is. A suit of armor might be a better representation—with dangerous chinks in it growing ever larger as the ozonosphere depletes and tears gaping holes over various parts of the body. (We'll look at this terrifying scenario later, when we examine a completely different process involving ozone—the day's usual buildup that results from increasing heat.)

Other than at noon in the tropics, light rays are always entering the atmosphere at an angle, greater or lesser depending upon latitude, season, and time of day. After a ray of light from the sun hits the outer stratum of air, what we see of it down here at the surface depends upon the physical attributes of each color of the spectrum. Visible light, as every rainbow demonstrates—and as recorded in the mnemonic abbreviation Roy G. Biv that schoolchildren learn—ranges from the red end of the spectrum, through gradations of orange, yellow, green, and blue, that indefinable dark blue called indigo, to the violet that arbitrates between blue and red and completes the circle on

11

artists' palettes. The words *infrared* and *ultraviolet* indicate that these forms of radiation are below (have longer wavelengths than) red and above (have shorter wavelengths than) violet.

Painters, or even children with crayons, quickly learn the frustrating quirk of physics that, while all the colors of light combine to produce white, all the opaque pigments combine to produce black. It is important to keep in mind, although difficult to remember when facing a coastal sunrise or a Paul Gauguin painting, that color is merely the animal brain's interpretation of variable wavelengths of electromagnetic radiation. And not even all animals see this way; moths perceive on flowers infrared images that are invisible to us. As Carl Sagan once remarked, "We might just as reasonably translate wavelengths of light into, say, heard tones rather than seen colors—but that's not how our senses evolved."

Imagine these light waves marching in a phalanx toward a ring of guards defending a castle, and it is not difficult to envision how some are likely to be repulsed and some to get through. Which are deflected and which permitted to continue moving forward depends upon the angle at which they enter the atmosphere—in other words, on how much air they must traverse between space and our eyes. One end of the color phalanx is less stubborn than the other. The blue end of the spectrum is scattered sooner than the red. As light makes its way downward through the atmosphere, the blue is deflected first, creating the effect of a blue sky surrounding the sun. Even the relatively thin layer of air directly above our heads is enough to scatter blue, so that it becomes the color we see overhead all day. Mars, in contrast, with under one percent of Earth's atmosphere at sea level, would normally have during its "daylight" hours a deep indigo sky.

Although this phenomenon is often just called *scattering*, physicists know it as *Rayleigh scattering*, after the English physicist John William Strutt, Lord Rayleigh, who first understood the process (and who also won the 1904 Nobel Prize for the discovery of argon). Scattering differs from *absorption*. As mentioned before, light waves are larger than the individual hundred-millionth-of-a-centimeter size of the simple molecules that compose most of the atmosphere. Such molecules, along with dust and other particulates, scatter light, but they don't actually absorb it. Absorption, the blocking of light waves, is caused by larger compounds such as nitrous oxides, those dangerous byproducts of combustion engines and industrial manufacture. Smog is brownish partially because it is genuinely absorbing sunlight.

Nor is our beloved blue sky an optical effect that you can observe *only* from the surface of Earth. It is visible from space. In 1961, Yuri Alekseyevich Gagarin, the pioneer Russian cosmonaut, became the first human being to see his native planet in its entirety. During his 108-minute flight, he was in orbit for less than an hour and a half, looking down from a height of 203 miles. But it was enough time and enough distance to leave Gagarin dazzled with the view. And one of the first experiences he later wrote about was the way that he could actually see the blue sky of Earth against the black of space: "The Earth has a very characteristic, very beautiful blue halo, which is seen well when you observe the horizon. There is a smooth color transition from tender blue, to blue, to dark blue and purple, and then to the completely black color of the sky." Only four years earlier, a retired American air force pilot, David Simons, had piloted a balloon through steadily thinning atmosphere to an altitude of more than nineteen miles, where he became the first eyewitness to report a dark purple sky. What a long and winding road from the footprints at Laetoli to the adventurous naked apes who climbed high enough—with the complex and long-term assistance of their fellows—to look down on home.

You don't have to be a cosmonaut to at least glimpse the sky's fade to darkness in space. Aboard an airplane at the usual cruising altitude of only a few miles, you can look out the window and see the lighter blue below, the darker blue beside you, and above you a slow fade toward an unfriendly-looking indigo. After this eerie moment, you may decide that the in-flight movie looks nicely distracting. No wonder blue is the color of most sky deities. As late as the mid-seventeenth century, in his *Art of Painting*, the Spanish artist Francisco Pacheco was insisting that Mary's cloak must always be portrayed as a celestial blue because no other color would do her justice. (Later, godless Nietzsche simply maintained that the lush blue afternoon skies in Turin helped him collect his thoughts.)

Red, in contrast, has a lower wavelength and moves more

lazily, plodding its slow way between obstructions. Only now at sunrise, and later at sunset, does it move through enough air, slicing horizontally through the atmosphere, to scatter and turn clouds into banks of red and orange. Physicists like to point out that you can witness this scattering effect with a simple experiment at home. Fill a fishbowl, or even a large glass pitcher, with water that contains just enough milk to slightly cloud it. Then shine a flashlight through it. The light by itself will appear yellow, like the sun, but observed through the milky water it looks reddish. Stay in the same position and have someone else move the flashlight around the back of the bowl, keeping its light aimed toward you. You will observe that the light changes all the way through the spectrum from reddish to bluish and back to reddish, as it angles through varying amounts of milky-water "atmosphere." (Naturally, different solutions will result in different colors and intensities.) Sky colors are created in the same way, by light bouncing off particles suspended in air instead of water. It makes sense that the more particles a beam of light encounters in its lower-angled flight through dense air at this time of day, the more dramatic the changes from midday's blue. The light is no more red upon its arrival at our atmosphere than it is blue at other times of day; it is produced by the milky air's filtering out of other colors.

Incidentally, without this scattering effect—these barriers to light shipping directly from its source to our eyes—even at sunrise the sky above us would be as black as the sky over the airless moon, blacker than Earth's darkest moonless night atop a mountain in the wilderness. Why a universe populated with stars would appear black instead of brilliantly glowing is another question entirely, and one that haunted astronomers' nights for centuries. We will address this enigma tonight.

In a rainbow the light's components are visible as parallel bands of color because suspended drops of moisture in the atmosphere serve as natural prisms and split the beams into their constituent hues. In the seventeenth century, Isaac Newton demonstrated this utterly counter-intuitive aspect of physics in one of the simpler experiments in the history of science—and still one of the more dazzling. He employed a prism to separate a beam of white light into its parts, and then another prism to reunite the segregated colors into a new beam of white light. Scientists have been infatuated with light ever since. It seems almost magical that the entire rainbow is hidden within the invisible rays of what we call light.

※

Morning is already moving quickly on the coast. A marsh that in darkness appeared as a primordial brainlike area of dark croakings and gurgles now reveals the byzantine pathways of streams wandering through it, as they reflect the lighter sky overhead. Dawn's screeches and croaks presage the seemingly lazy flapping of great heron wings across the marsh. Then competitive symphonies begin, glee clubs of declaration and unification, rising and falling cadences like the later cicada chorus that will accompany the heat of midday.

There seems to be available at dawn a richer and more pure pink than sunset can afford. Clouds quickly turn from black to dark gray through the subtle gradations of blued grays and grayed blues that indicate how insubstantial, how chameleonic, clouds are—which is why it is always a moment of pure fantasy to fly into a cloud in an airplane. A single cloud, depending upon its successive positions in the parade, goes through quick progressions of blues and pinks. Just before the arrival of the sun itself, many clouds appear pink below and blue-gray or purplish above. The advance pinks brighten to red-gold, to a golden white along cloud edges. Never do they show their true colors; like royal attendants they reflect the mood of the ruler.

Part of the drama of dawn comes from the different speeds with which clouds move across the sky, like stage sets shifting to left or right to reveal another set behind, ready for the next act in the daily drama. Finally there it is, the first piercing glimpse of the sun, climbing above the horizon. The avian orchestra grows stronger. Now that this long-awaited personage has arrived, you must look away. Anyone present at the daily re-creation of the world knows that the sun is not a nuclear fire millions of miles away. As dawn reminds us every time we go out to meet it, the sun is a god. However "mistaken" the old fables may be in their stories of how things came to be the way they are, they intuitively perceive the dependent nature of our relationships with the natural world, and they express them in the most natural mode available to any creature who experiences life in a linear progression—narrative.

Turn from the rising sun and you find its luster first touching the tops of other trees nearby. Why is this golden moment so heartening? Some of us rise at dawn simply to be present at this daily show. Something childlike leaps in us and cries out, "The sun is up!" And once again we have that ancient yet ever-new phenomenon, beloved of poets and pedagogues, tarnished by greeting cards and commencement speeches, yet nonetheless eagerly awaited by commuters, bird-watchers, poets, children, and survivors of tragedy: a new day.

I say, the sun is a most glorious sight:
*   I've seen him rise full oft, indeed of late*
I have sat up on purpose all the night,
*   Which hastens, as physicians say, one's fate;*
And so all ye, who would be in the right
*   In health and purse, begin your day to date*
From daybreak, and when coffin'd at four-score
*   Engrave upon the plate, you rose at four.*

—Lord Byron, *Don Juan*

Smiley

The rising sun shows up frequently in children's drawings, and often it bears a visage like the supposed face of the man in the moon. Interestingly, children are not the only ones who have portrayed the sun this way, even though we can't look at its disk directly, much less perceive any features on its surface (or at any rate on the gaseous hell that passes for its surface from this distance). An illuminated manuscript page from India, probably from the eighteenth century, portrays Agni, a deity of ritual and fire, genuflecting before an altar, above which floats—à la the Great and Terrible Oz—a large solar face, complete with dimpled chin, nasolabial folds, slightly bloodshot eyes, and even ears (on the inside of the circle). The flag of Uruguay bears a full golden sun, complete with face. Surely the once ubiquitous smiley face is yellow partially because it has been influenced by images of the sun. Among the countless representations of the Buddha, there are both moon-faced and sun-faced Buddhas; supposedly the latter have lived longer in the light of wisdom.

Another form of cartoon shorthand that shows up in pictures of the sun is the representation of light rays spiking outward from it. This image is also not limited to the artwork of schoolchildren. Consider the huge solar disk above the undulating fields and hills in Van Gogh's drawings. The Dutch artist emulated the wavy lines and forceful outlines of Hokusai and other Japanese printmakers whose work helped turn him away from the conventions of Western painting. In his drawings Van Gogh suggests the sun's radiance, both its light and heat, by concentric circles of dashed lines. Similar visual shorthand brings to life his painted suns. *The Sower* is dominated by the sovereign climbing above the horizon, showering Danaë-worthy gold in every direction. Van Gogh's sunflowers appear as stylized suns rayed with golden petals, and even the hanging overhead lights in *The Night Café* radiate solar rays. Although derided by Van Gogh's contemporaries, such representations became in the twentieth century standard visual shorthand everywhere from animated cartoons to advertisements for air-conditioned relief from summer heat.

I used to be amused by those men who get to work at six-thirty, "bright and early"—but they're right: you want to be doing things when the world is still quiet; the quietness and uncrowdedness is your fuel.

—Nicholson Baker, *A Box of Matches*

A Perfect Story

"Every day . . . is an artistic whole," wrote the essayist Christopher Morley early in the twentieth century; "it comes out of nothing and goes back to nothing, like a perfect story." Many writers have found the natural cycle and narrative drive of a single day irresistible. Twenty-four centuries before Morley, in his *Poetics*, Aristotle argued on behalf of the dramatic unity of place and time. Many novels condense their narrative into a single day; Moacyr Scliar's *Strange Nation of Rafael Mendes* and Yasmina Reza's *Adam Haberberg* are two recent examples. Aleksandr Solzhenitsyn needed only one day in the life of Ivan Denisovich to dramatize the horrors of Soviet labor camps. Thornton Wilder's play *Our Town* unfolds in a single day, observed by the avuncular Stage Manager as the inhabitants of Grover's Corners awaken, go through their daily routines, and settle in for the night. Probably the most famous novel that unfolds during a single day is James Joyce's *Ulysses*. Although patterning his work after

Homer's *Odyssey*, which has a timescale of ten years, Joyce wove all the monologues and descriptions and jokes and puns of *Ulysses* around one day in Dublin—16 June 1904, ever after called Bloomsday by Joyceans and tourist-luring Dubliners.

The English novelist John Lanchester permits the reader to accompany the titular Mr. Phillips through a day in which, because he has lost his job (become "redundant," in that terrifying British phrase), he simply wanders around London observing the lives of his fellow creatures—and primarily thinking about sex. In elegant and meticulous prose, Lanchester follows the rhythms of the city throughout the day: the steadily changing light, the growing and then declining heat, morning and afternoon lulls that bookend the midday feeding frenzy.

In the 1920s, while she composed her novel *Mrs. Dalloway*, Virginia Woolf was reading Euripides and Sophocles, the same Greek tragedians that Aristotle invoked in his argument for the unity of place and time. These two elements could hardly be more dramatically wed than in Woolf's magnificent story of a single day in the life of Clarissa Dalloway and poor mad Septimus Smith. Several years later, Woolf employed an extended day-as-lifetime metaphor in her novel *The Waves.* Six narrators alternate, overlapping their memories of each other, in sometimes awkwardly stylized diction, beginning in childhood and moving forward through maturity and into old age. Yet Woolf has these characters recount their lives in the context of a single day. The book opens with dawn over the ocean, as waves roll in to the shore. Light touches the water and then the land: "The sun sharpened the walls of the house, and rested like the tip of a fan upon a white blind and made a blue fingerprint of shadow under the leaf by the bedroom window." Periodically Woolf returns to the sun, its climb toward zenith, its slow fall westward.

The narrative pull of the day appeals to more than fiction writers. During the last couple of decades of the twentieth century, a series of popular books sent photographers to document the variety of experiences in a particular area—from Las Vegas to Australia—in a single day. Inevitably the symbolic value of different times of day shows up in film and television, from *The Dawn Patrol* and *High Noon* to *After Dark, My Sweet* and *The Twilight Zone* and *Midnight Cowboy.* Because we love to impose a narrative arc on a period of time, occasionally we take this tendency even further. As John Updike observed about the end of the twentieth century, "Centuries feel bumptious in the beginning, thick in the middle, and sickly at the end." Hence our preoccupation with fin-de-siècle culture, which—at least when the term referred to the end of the nineteenth century—usually meant decadent and irresolute. Yet no period of time is more artificial than a century, and relatively few human lives witness one in its entirety. The alternation of daytime and nighttime, however, we have been witnessing since birth; it has been dominating the rhythms of our own bodies since long before our kind first evolved.

Chariot

The Endless Array

Most of us are familiar with the words of Eleanor Farjeon, although we're likely to remember the Cat Stevens version: "Morning has broken, like the first morning. . . ." Each dawn does seem magically able to recall some kind of mythic, primordial daybreak. As the sky gradually lightens you can sit in your own backyard and watch the world returning. Subtleties of tone emerge surprisingly quickly. At first the trees are as massed and indistinct as clouds. Slowly they resolve from a mountainous accumulation of silhouettes to layered tones, the outer leaves now backlit, almost translucent if up close, those farther back or deeper within the crown layered into opacity. In the low light, new yellow-green growth on hedge and shrub is visible sooner than dark older leaves. Pale blossoms identify themselves before the darker leaves behind declare individuality with field-guide silhouettes against the lighter sky. "Dawn," wrote the Chinese poet Meng Hao-jan in the eighth century, "ignites the endless array of things." It returns the world to us.

Dawn in winter and summer are entirely different phenomena. In winter the light comes more quickly to a harsher landscape, no softness, no lazy caress of foliage. Sounds are more stark, less birdsong, no leaves cushioning the rattle of a passing commuter train. Bare deciduous trees are reduced to an India-ink sketch, although the dry-brush angularities of conifers remain approximately the same throughout the year. Winter days also dawn with no reassurance that warmth will make at least some part of the day pleasurable, so they are not as satisfying in their arrival.

Warm weather brings the finest dawn, as trees bud in early spring and young leaves blur the sketch like a first wash of watercolor. The chorus of birds, the breakfast errands of squirrels, a rabbit watchfully ruminating in a corner of the lawn—these morning glimpses remind us that other creatures share our day/night bias. One of the themes from bedtime stories that small children find comforting is the idea that animals also prepare for bed—in nest or burrow—and sleep and awaken to a new day. At dawn we are reminded that other creatures are likely to spend their day much the same way we will: foraging and squabbling, protecting their young and watching for predators, tidying the burrow.

In most places around the world, at least in warmer months, dawn provides as much to listen to as to look at. A sufficiently mimetic composer—a mynah, for example, or its talkative cousin the starling—might collage a symphony of the day through careful juxtaposition of sounds. Begin with the avian chorus that sounds to us like a collective hosanna.

"It was five o'clock and the birds were singing with crazy joy," writes Iris Murdoch in one of her novels. It does sound irresistibly joyful to us—usually. If we sleep with a window open and hear the birds stirring hours before daylight, however, they sound about as charming as a single insomniac cricket who begins practicing scales as soon as you turn out the light. Ornithologists remind us that this dawn hubbub sorts into predictable categories like other animal sounds, including human ones: promise, swagger, threat. It's nice to imagine that there's some pleasure mixed in. In eastern North America these sounds up close become the chuckle of robins on the lawn and squabbling jays overhead, the gimme chirps of baby orioles swaying in their pendant nest. Pigeons, those omnipresent rock doves that have colonized the office buildings of Seattle and Amsterdam the way their ancestors colonized the cliffs of Gibraltar, call and flutter on windowsills and rooftops.

But birds are only the most obvious noise audible to the early riser. Inside the neighbor's house a dog's request for an outing is muffled. Floors creak as feet walk on them for the first time in hours. There is no longer the rattle of glass milk bottles being set on a doorstep, but a listener hears the early-morning thud of a rolled newspaper and eventually the opening front door, the scuff of slippers on a still-dewy sidewalk. Mix in the distant-waterfall noise of traffic gradually increasing from a sigh to a roar. Beyond the hedge you hear the distinctive jangle of the same dog's collar as he walks a yawning human and reads the olfactory bulletin board. As the dew dries, disheveled bees emerge into sunlight and strafe clover blossoms with a pensive buzz. On ordinary days, those without thunderstorm or hurricane, only at dawn does nature achieve aural dominance in the city. In urban areas, the quieting of the birds accompanies increasing traffic noise, honking horns, sirens. Baby pigeons coo in their nest on the underside of a bridge overpass, their calls and the flutter of their parents' wings drowned out by the rattle of delivery trucks overhead. It is an unpredictable new day in their lives as well as ours.

※

In Greek mythology, Eos was the personification of dawn. (She was Aurora to the Romans.) The daughter of Titans Hyperion and Theia, Eos is in some accounts the sister of Helios and Selene, the sun and moon. She is also the mother of the winds, and in a few stories the mother of Phaethon, whom you will meet shortly. When the chariot of the sun is ready to depart heaven, her rosy fingers open the gates.

Judging by her innumerable amours in ancient mythology, she must have taken the rest of the day off for romance, but her daily wake-up signal was enough to ensure linguistic immortality. Primordial stones are called eoliths. Charles Lyell, the Victorian geologist, designated the second great epoch of the Tertiary period (the so-called Age of Mammals) the Eocene. Before it was even identified from fossils, Lyell's colleague Thomas Huxley named a hypothetical five-toed ancestor of our familiar one-toed horse *Eohippus*. The identification of the long-awaited fossil provided some of the first hard evidence of the triumph of evolutionary theory. Because

Eohippus is so important in the history of science, it seems a shame that, for taxonomic clarity, the genus of the dawn horse has been reclassified as the inelegant *Hyracotherium*. A talented draftsman, and a man who could hardly think at all without waking his sense of humor, Huxley once dashed off a cartoon of the earliest horse, galloping on five-toed feet, with a hairy protohuman astride it. Under the human figure he scribbled "Eohomo," "dawn man," a joking term that would reappear a half century later as *Eoanthropus,* the short-lived name for that legendary impostor Piltdown Man.

※

So many human beings awaking anew each day, so many faces turned like sunflowers toward the east, honoring the dawn in religion and literature and film. Muslims begin the fast of Ramadan just before dawn; a small meal is permissible only if eaten before light appears in the east. One of Homer's stylistic tics that is immediately apparent to readers of the *Odyssey* is that a sunrise must be heralded by the rosy fingers of dawn. Thoreau couldn't resist sermonizing: "Only that day dawns to which we are aware." The symbolic new beginning of dawn is as irresistible to filmmakers as it is to writers. It provides powerful visual metaphors (and blissful cinematography) in such films as F. W. Murnau's silent *Sunrise*, and even in the light of a new black-and-white day washing over Baghdad-on-the-Hudson in Woody Allen's *Manhattan*. Naturally this time of day, like sunset and birth and winter and so many other natural phenomena, has been forced to dance for kitsch and commerce. Has any politician ever resisted the temptation to inflate a speech with the phrase "the dawn of a new day"?

Like every other experience in life, dawn inspires conflicting emotions, no matter what its established symbolism in myth and art. We may imbibe the traditional symbols, but we respond to them in our own way. It is interesting to contrast two views of dawn by two literary-minded French writers whose almost back-to-back lives stretch from the Romantic 1820s to the middle of the twentieth century. That great noticer of natural phenomena Colette paid particular attention to this time of day, and even honored it with the name of one of her books. The title *La Naissance du jour* (Dawn or, literally, The Birth of Day), reminds us that each day starts young, that every morning is new, that a renaissance is etymologically a rebirth. In her 1929 novel *Sido*—actually more *mémoire* than *roman*—she says that during her childhood her mother would grant her the dawn as reward for good behavior. She would wake the young Colette at 3:30 a.m. Slipping outdoors into the still darkness before sunrise, Colette would disappear into the world of her senses: "It was on that road and at that hour that I first became aware of my own self, experienced an inexpressible state of grace, and felt one with the first breath of air that stirred, the first bird, and the sun so newly born that it still looked not quite round."

Baudelaire did not respond to early morning with quite the same rapture. Consider his poem "Le Crépuscule du matin" (Morning Twilight), from *Les Fleurs du mal*, first published in 1857. Baudelaire's vision of this time of day, as any reader of his can predict, is vivid and concise but hardly rosy.

The poem is populated by images that don't show up in your average ode to daybreak: starving beggars blowing on their fires, prostitutes sleeping with their decadent mouths open, and the indigent dying in charity wards. The very air is full of shudders. It is the hour, writes Baudelaire, "when women's labor pains grow more cruel." Dawn shivers in its pink and green raiment. And the cackle of a distant and apparently consumptive cock sounds not carpe diem, not Thoreau's clarion to wakefulness, but instead rents the fog "like a sob stifled by a bloody froth." Thanks to some kind of cosmic mismanagement that clearly depresses M. Baudelaire, Earth has once again turned toward the sun.

For the rest of my life I will reflect on what light is.
— Albert Einstein

A Year of Light

Before we look more at the world around us, let's consider what is enabling us to do the looking. The phenomenon that we call "light"—omnipresent, mostly ignored, and deeply weird—is only the visible portion of the electromagnetic spectrum. Electromagnetism, in turn, is only one form of energy. "Science shows us that the visible world is neither matter nor spirit," wrote the American physicist Heinz Pagels; "the visible world is the invisible organization of energy."

Other kinds of electromagnetic radiation, including X-rays and radio waves, also exist in the form of these oscillating bundles of energy. The sun is showering radiation on us all day long—some in the form of visible light, some not—and so are the groves of television and radio towers on higher ground around the world. A number of other natural phenomena produce electromagnetic radiation. As radio-telescopes document every day, our planet collects its share of radio waves from interstellar clouds. We receive an unremitting microwave bombardment from cosmic background radiation, the residual buzz of energy that seems to have been left over from the Big Bang. The astrophysicist George Gamow first proposed the idea that this radiation from outer space might result from residual heat from our universe's initial explosion. He argued that the photons of the original bang would have begun as visible light but would have gradually wound down to the lower radiation frequencies, which would explain the presence of microwave radiation from space.

In his photoelectric theory of 1905, Einstein explained that light consists of photons, a minuscule indivisible unit of electromagnetic radiation. He was applying to light Max Planck's revelation from five years earlier, that energy exists in discrete packets called quanta—which did for energy what atomic theory did for matter, establishing its basic unit. (Actually physicists now consider quarks, particles thought to have been produced during the Big Bang, to be an even smaller and more primal unit than atoms.) The once-shocking notion that light can be thought of as either wave or particle, because it behaves differently under different conditions, is now a given, like so many other hard-won discoveries about nature.

For all practical purposes, to those of us whose everyday lives do not include particle accelerators or spectroscopic analysis of distant galaxies, light moves instantaneously. Across the tiny distances of our planet, light is a genie, arriving anywhere the moment that it thinks of going there. Because it will come up elsewhere in this book, however, and because—out of some crazy arbitrariness at the core of creation—the speed of light appears to be the speed limit of the universe, let's consider it for a moment.

In the near vacuum of space, light travels at a speed of— or at least it has been officially defined as—299,792,458 meters per second, usually rounded to 300,000 kilometers, which is almost 186,300 miles. This evening we will return to the speed of light because of its ability to convey to us the way that the universe looked millions of years ago. But these matters come up only when we look beyond our own relatively nearby sun, so we will wait until nightfall and the stars come out, broadcasting their antique light across emptiness.

Until we have to place telephone calls to resorts or retirement communities on the moon, the speed of light is not very important to most of us. As you may remember from television news conversations with the Apollo astronauts, there is a disconcerting lag time between speaking into a microphone on Earth, the call's reception on the moon, and transmission and receipt of the astronaut's reply. In the vacuum of space, all electromagnetic waves, radio or light or whatever, travel at the same speed. The moon averages 239,000 miles (384,000 kilometers) from us. If you divide the distance to the moon by the speed of light, you find that it takes about one and a third seconds for light (or radio waves) to travel between us and our satellite, creating a conversational delay of roughly three seconds between question and reply. Soon this interval will become an accepted part of our daily chats with people on the moon.

Imagine pushing the moon ever farther away, until the one and a third seconds becomes eight minutes and more, and you will comprehend the relative distances of the moon that orbits us and the star that we in turn loyally circle. Then keep moving outward. Look again at the figure showing how far light travels in a single second. Multiply this figure by 60 seconds in a minute, 60 minutes in an hour, 24 hours in a day, and 365 days in a year, and you get some idea of the incomprehensibly vast distances to which astronomers casually refer with the term *light-year*. (Mark Twain once performed this calculation. Hours later, during the author's daily billiard games, a friend found him still dazzled by the magnitude of such numbers, "trying to compass them and to grasp their gigantic import.") Multiply the total by the number of years in your lifetime so far, and then turn to a sky guide and look for an astronomical body that is roughly as many light-years away as you are years old. Go out into the evening and find the star. Its patient photons will immediately enter your eyes and be processed by the photoreceptors in your retina. You are responding to light that left its faraway nuclear source at about the time that you were born. Ever since then, as you played and learned and stumbled and slept and graduated and dated, as your body and mind accumulated experience, these photons were moving through space. And at this moment, as you raise your eyes from the star guide to the sky, you are ready to meet these time travelers and welcome them to your home planet.

Down to Earth

The mean distance between our planet and its star is roughly 93 million miles or 150 million kilometers. Let's examine this number briefly. Just as our twin pentadactyl (five-fingered) mammalian forepaws inspired the decimal system—and consequently such artificialities as centuries and millennia—so have we promoted other chance numerical arrangements to measuring systems. How else could measurement begin? Using the distance from Earth to sun, astronomers divide space into multiples of a basic Astronomical Unit (AU) of 150 million kilometers, just as they measure planetary mass in multiples or percentages of Earth's. By such reckoning, Saturn is 9.54 AU from the sun, Pluto averages 40, Mercury only about .3.

We must remember that Earth's orbit is elliptical, and that the AU is an average between *aphelion*, Earth's farther distance from the sun, and *perihelion*, the nearer distance. It's true that the words *apogee* and *perigee* describe farthest and closest distances between any body and the body it is orbiting, but the terms above refer specifically to orbiting the sun—and besides, they honor Helios, one of the Greek solar deities.

The AU is a provincial unit of measure, as quaint as pacing off property boundaries or the way that in medieval England the yard supposedly originated as the distance between King Edgar's nose and the outstretched tip of his middle finger. But it is a unit worth remembering as you navigate the super-scripted zeroes of cosmic distances. Ever prodding nature to vouchsafe a few more intimacies, scientists employ the AU in various ways, including an equation deducing a planet's period of orbit from its distance from the sun. This equation expresses the third of the brilliant laws of planetary motion formulated by the German astronomer Johannes Kepler in the early seventeenth century: $P^2 = a^3$, in which P represents a planet's period of revolution around the sun (measured in terrestrial years) and a the planet's distance from the sun (measured in AU). This equation is not merely scientific arcana. It demonstrates yet another way that we have responded to Earth's primordial relationship with its parental star, expressed via our usual—indeed, inevitable—habit of describing the distant in terms of the local. By defining the motion of other planets in terrestrial terms, Kepler's third law distills as much about human computation as it does about the mathematical harmonies of space.

Even a unit the size of the AU, however, remains useful only in the immediate neighborhood. Beyond our own solar system, measuring in piddling little AUs would be like count-ing a human lifetime in seconds instead of years. The result would be numerical chaos. How many seconds old are you? Yet the next step up, a light-year, is also based upon the sun and Earth. It measures the distance that light travels through the near vacuum of space in one terrestrial year, although the duration of our year means nothing to other planets or stars. The speed of light, as noted a little while ago, is roughly 298,000 kilometers per second. Should its physics permit such boastful displays, a photon could circle eight loops around Earth in the amount of time it takes you to say aloud "a single second."

These matters are not abstract or distant from daily life; our relationship with the sun is deeply personal. Only a vivid realization of how fairy-quick light is can demonstrate the impossible distances between us and even our familiar astro-nomical neighbors. At this amazing speed, it still takes photons more than eight minutes to travel one astronomical unit—to rocket from their nuclear-powered source, traverse millions of miles of emptiness, ricochet from pollen grain to dust mote down through Earth's atmosphere, squeeze between miniblinds, and arrive in time to gently backlight the downy hairs on your daughter's earlobe just before you wake her for school.

Yesternight the sun went hence,
And yet is here today.

—John Donne

Sun Sister and Moon Brother

As our story progresses, we will address the different phenomena behind—and the various speculative rationales
for—the unique attributes of each time of day: dawn, morning, noon, afternoon, sunset, twilight, nighttime. But for now
let us attend to the most basic point of all. All over the world,
long before we understood the astronomy involved, we told
ourselves Just-So stories to answer this reasonable question:
"Why does daytime alternate with nighttime?"

One curious justification for the rhythm of light and dark
appears in a myth of the Yuchi, a North American tribe
native to a region now in eastern Georgia and western South
Carolina. In the early days of the world even Sun and Moon
have animal form and can dance upon the rainbow. But the
decision to create the earth demands light to banish the
omnipresent darkness, and various animals try to illuminate
the murky new world. Both Glowworm and Star fail to produce enough light; although brighter, Moon is also not quite
up to the task. Only Sun succeeds. She climbs up to zenith
and then debates whether to continue toward the western
horizon and disappear for the night or remain overhead and
provide uninterrupted daylight. She appeals to her fellow
creatures and they put the question to a vote. Finally, in a
most unanimalian blush of modesty, the animals decide to
alternate daylight and darkness so that nighttime will offer
them privacy for sexual intercourse.

In some myths Sun Sister and Moon Brother chase each
other across the sky. Some Hindus believed that Siva formed

day by opening and night by closing his eyes. Natives of the Arctic coasts of Canada once attributed day and night to Hare and Fox. In the dark early time, Hare beseeches the gods for light so that he may see to forage, but Fox begs to remain cloaked in the darkness that veils his own hunting. Hare wins the argument and is granted light for part of each day. A widespread story around the world—the idea of stealing fire, as in the Prometheus myth—appears now and then as the theft of daylight itself, as when the Raven of Kwakiutl myth captures the sun in a box and brings it down to earth.

As such fables demonstrate, the daily astronomical rhythm of light and dark is ingrained in our consciousness. Usually we don't even notice it. When we employ it as part of an automatic linking of unrelated phenomena, we are performing the sort of binary thinking that psychologists call dual symbolic classification. Our symbol-hungry brain observes the seeming opposition between light and dark and quickly leaps from the actual experience—awareness versus unconsciousness, the relative safety of daytime versus the unseen dangers of night—into an opposing dichotomy that pervades the arts and even casual speech. For thousands of years we have equated daylight and the sun that provides it with birth, summer, reason, sight, honesty, and virtue; in contrast nighttime equals death, winter, irrationality, blindness, falsehood, and evil. Such associations cling to the sun and the moon even now. Day is the time for honest work and night for villainy. The same sort of pairing shows up, for example, in the Bible, when Jesus predicts that God will gather the righteous on his right and the damned on his left, an ancient equation between majority and virtue that Michelangelo depicts in his Judgment scene on the Sistine ceiling.

Too often most of us default to duality symbolism when instead we could use some healthy syncretism. Originally derived from a Greek term meaning a union of Cretans, and then a unified front against a common enemy, syncretism has come to refer in religion and philosophy to the reconciliation of opposites. To syncretize is to form a confederation, as in a coalition government. In psychology syncretism refers to the merging of contrasting elements into a single image, which would seem to make it also the study of metaphor, and in grammar it means the confluence of linguistic tributaries into a single channel. Syncretism opposes the us-versus-them attitude, the all-or-nothing approach that fuels, say, politics, sports, and fundamentalist religion. It is the philosophy expressed in the *Tao Te Ching*. We know beauty only because there is ugliness and we know goodness only because there is evil; long and short, high and low are defined by their opposite; "front and back follow one another."

An illuminating example of syncretism occurs in the Eastern idea of the yin and yang, the balance between complementary principles of being. *Yin,* the female power, represents passivity, the earth, cold, darkness, night—the primordial darkness that is not evil in itself because from its potent fertility all else is born. The male *yang* is the opposite: action, the heavens, warmth, light, daytime. This is hardly a feminist cosmogony, but for most people the now common symbol seems to have lost its gender bias. Confucian and Taoist commentators insist that the two forces are equal in power, alternately dominant or submissive and achieving an ultimate harmony.

The *Ta ki,* the familiar visual representation of yin/yang, embodies the balanced tension between the two by bisecting the encircling cosmos with a sigmoid (S-shaped) line rather than a mere diameter. In contrasting shades, it portrays equal opposites embracing each other, interdependent, uniting to form the harmonious circle of the cosmos, each with the seed of the other like an eye in the head of its fetal form. In Taoism, this alternating cycle governs the rhythms of nature, from the largest we observe to the smallest.

Spin

With mythological rationales and cosmic harmonies vivid in the mind's eye, we ought to examine the *real* origin of the phenomenon that not only creates day and night but also subtly influences almost every aspect of it, from weather to circadian rhythms in plants and animals. Where did the day really begin?

During the larceny trial of the Knave of Hearts, the royal judge looks down from his bench at Alice in the witness box and instructs her, "Begin at the beginning, and go on till you come to the end: then stop." The King's command, which sounds like fine advice for courtroom witnesses and freshman writing students, is not always easy to follow. Where do we begin the story of the day? Like so many questions about nature and science, this one reveals long-running interactions between many different natural processes. "When we try to pick out anything by itself," as John Muir famously remarked, "we find it hitched to everything else in the Universe." No topic is larger or more influential, more firmly hitched to a bewildering array of phenomena, than our planet's daily spin on its tilted axis.

But let us try to follow the King's advice. We might begin by trying to rewind time like a movie, not just for a few frames but to the beginning of the film. Watch in your mental studio as the movie plays in reverse. Earlier today, just ahead of dawn, we imagined the long era before artificial lighting, but now let's start there and go much further back. Urban crowds scatter into the countryside and cities sink into the earth. Our ancestors take off their clothes and cluster in tropical Africa; they grow shorter, hairier. Gradually the great capering pageant of mammals—giraffes and chimpanzees, saxophonists, acrobats—merges into a few nimble forms dodging the tree-trunk legs of giant lizards. The sea and sky are full of monsters. At first mountains grow rather than erode, but finally they sink into the sea. The Earth warms; water evaporates. The continents join together like pieces of a jigsaw puzzle, the eastern shoulder of South America tucking cozily into the western curve of Africa. Finally we are so far back in time that the surface of Earth is clothed in toxic vapors, then molten, then ceasing to be our planet and becoming more like other planets in our solar system: a slowly coalescing ball of minerals, inimical to life. If what we think of as a "day" had a true genesis, it must have been when the proto-Earth first began turning on its axis.

Yet, according to the scientific consensus early in the

twenty-first century, the materials that eventually composed the primordial Earth were spinning *before* they became what we now think of as a planet. Over the centuries, astronomers have proposed many theories of how the solar system came into existence. It was once suggested, not unreasonably, that the planets might have formed when the sun's gravitational pull attracted solid material as it journeyed through our region of space—fragments of comets and other debris, the cosmic equivalent of driftwood. Early in the twentieth century the English astronomer James Jeans theorized that the planets might have formed after a comet or another star collided with the sun, forcing the ejection of balls of hot gas that then coalesced into planets. Most astronomers, however, now bet on what they call the *nebular theory*, the idea that the young sun was attended by a disk of gas and dust, from which planets eventually conglomerated. Now that we have begun to peer more closely into the private affairs of distant galaxies, thanks to the Hubble Space Telescope and other innovations, we can see that many younger stars, such as those observed in the Orion Nebula, are indeed encircled by a ring or disk of gas and dust.

The four inner planets of the solar system—Mercury, Venus, Earth, and Mars—are usually described as rocky to distinguish them from the gaseous outer planets—Jupiter, Saturn, Uranus, and Neptune. Pluto also has a hard stone surface, but its recent demotion from planetary status bans it from our list. (Of course, although we now speak casually of "the planets," for a long time Earth was not considered one of them; it was the stationary constant underneath the dancing heavens.) One scenario that scientists propose is that, in the early period of what would become the solar system, the inner planets began with the slow accumulation of microscopic grains of dust from the nebular cloud surrounding the sun. Heating melted the grains into larger particles of a millimeter or more in diameter, and these in turn accreted into boulder-size planetesimals. Not until they reached a diameter of a kilometer or so did the planetesimals each acquire a gravitational pull captivating enough to attract other planetesimals and merge into ever larger corporations of rocky material. Not surprisingly, the duration of these planet-forming events is difficult to determine, and estimates vary from ten to hundreds of millions of years.

Probably the various rocky planets began in a similar manner. As a consequence of their unique situations, however—each a different distance from the sun, each with a separate physical history of impacts from space, and so on—each developed along its own path. The poisonous clouds of Venus, the arid sand of Mercury, the gigantic ongoing storm that is the wandering Red Spot of Jupiter: alternative histories of our own planet are not difficult to imagine. Neither poets nor creationists can get over the wonderful oddity that Earth developed water and eventually began to metamorphose hydrogen and oxygen into kelp and red-winged blackbirds. But it is surely wonderful enough that long before life evolved here, even before the planet became a planet, it was turning on the cosmic wheel, alternately light and dark like yin and yang, and clearly attuned to cosmic harmonies of which we ourselves are normally unaware.

Apollo Bright and Pure

In mythology, such cosmic rhythms intuitively find expression in human form. Light and darkness, love and hatred, life and death—embodied, they struggle in hand-to-hand combat, beget offspring, reconcile. They carry the world and hang the stars. They enact physical processes and symbolic connections in forms that we can recognize and with which we can sympathize.

No solar deity more fully embodies notions of proud daylight than Apollo. Although he seems originally to have come from elsewhere, by the time of the *Odyssey* and *Iliad* Apollo was established as the son of Zeus, the self-indulgent don of the quarrelsome and licentious clan of Greek gods. The god of music, he also broadcast his father's will via oracles, monitored civil and religious policies, and inspired and absolved guilt. Epithets for Apollo, honoring attributes or denoting regional cults, included Nomios (herdsman), Alexikakos (averter of evil), and Delphinius (for one of his totems, the dolphin).

The epithet that concerns us here, Phoebus, was a forename meaning both "bright" and "pure." The name Phoebus Apollo honored the god's relationship with the sun, which seems to have been a later addition to his other powers and duties. Sometimes Apollo is described as the personification of the sun's spiritual grandeur, its omnipresence and presumed omniscience; one of his traits was that he could observe or communicate or punish from afar, and even his fellow deities trembled before him. An example of Apollo's figurative farseeing is that he bestowed upon Cassandra the ability to predict the future; but when she dared to spurn his love, he cursed her with true prophecies that no one believed—including her forewarning to the Trojans, which they ignored, about soldiers hiding inside the gift horse from the Greeks.

Helios, in turn, was "merely" an incarnation of the sun as an astronomical entity, a physical fire in the sky. (Leave it to our imagination to declare that the properties of something were nobler in some fanciful hierarchy than the physical reality that produced those properties.) Later myths conflated or confused the various Greek and Roman sun gods. Contemporary language has kept these otherwise retired divinities employed to describe solar matters. *Apollonian* describes the supposedly reasoned and honorable aspects of the day, opposite the uninhibited revelry of nighttime and drunkenness (also dubbed with a mythological name, *Dionysian*). *Heliotropic* is a more practical word, referring to plants that automatically turn their growth toward the sun as it arcs overhead, and *heliocentric* describes the sun-centered view of our planetary system that replaced the earlier geocentric view. The American Apollo space program, which landed human beings on the moon, was ill named; the Soviets had a more apt designation in their *Lunik* (Luna) craft.

While his name adorns spacecraft in museums, some of Apollo's role as a solar deity has been taken up elsewhere. Consider one contemporary manifestation of mythology— Superman. Despite his resemblance to Earthlings (and his impressive ability to disguise himself with only a pair of glasses), Superman is from another world. He is born as

Kal-El on the planet Krypton, whence his father Jor-El rockets him to Earth because Krypton is about to explode. Kal-El acquires superpowers when exposed to our sun's yellow light. According to a breathless aside in an early Superman comic, "Krypton's Jupiter-like size and red sun kept the Kryptonian race weak, while on Earth Krypton's last son is the mightiest of all!" Solar from his very genesis, Superman— the very antonym of the dark and Dionysian Batman—is Apollonian to the core.

Some etymologists trace the name Apollo to the Greek *apollynai*, "to destroy," and also link it to the destructive ability of lions. In Greek mythology the sun was linked with the zodiacal sign of Leo. This association was not unique to the Greeks; the lion was also a symbol of Mithra, the pre-Zoroastrian sun god of Iran. In some places a young lion represented sunrise, a mature one noon, an aging one sunset. The Egyptians credited this beast with power over weather because the Nile flooded during the season in which the sun rose from the constellation of the lion. Aside from its vigor and tawny color, the animal's mane alone might have inspired such analogous thinking. In many ancient cultures long hair was reserved for kings, because—seemingly independent of and outlasting the rest of the body—hair was associated with power. Usually sun gods were portrayed with long hair, because hair radiates outward from the face in the way that rays emerge from the round sun.

Sunny Apollo was brother of the moon goddess Artemis— a.k.a. Diana, Selene, Phoebe, and Luna. In many mythologies, solar and lunar deities are either siblings or married. In some, of course, they are violently opposed, as in widespread images from Asia and Africa portraying the solar lion ripping out the throat of the lunar bull. Many cultures assigned the stronger light a male role and its pale nighttime mirror a lesser female identity. No matter how a culture explained the relationship between sun and moon, these essential lights began as the principal deities of day and night.

The Son of Fire and Water

Apollo features in a popular myth that involves a break in the cosmic order upon which so much of our thinking about daytime has relied. The tragic tale of Apollo and his illegitimate son Phaethon is a story that embodies both the mythological view of the cosmos and the daily relationship between the sun and the earth below it.

Like many characters' names in mythology—or for that matter in Dickens or Wodehouse—the name Phaethon (which has three syllables, with the accent on the first) expresses the character's essential characteristics. The Greek *Phaethon*, which literally meant "Shiner," derived from the present participle of *phaethein*, "to shine," and ultimately from the root *phos*, "light," the root of such words as *phosphorescence*. Phaethon's history appears in various sources, but the best, as usual, is in that bottomless well of stirring adventure and delicious writing, Ovid's *Metamorphoses*. Ovid makes the story exciting, archetypal, and movingly human all at once.

Phaethon is the son of a broken home. His mother is the sea nymph Clymene, with whom he lives on Earth; his father, or at least so his mother claims, is Apollo—or Phoebus or Helios in various accounts. Because of this heritage Phaethon was sometimes called the Son of Fire and Water. A typically insecure teenager, Phaethon brags about his famous father. His schoolmates are skeptical. They suggest that Phaethon's mother has manufactured the divine fatherhood story to cover her indiscretions (as both scholars and comedians have suggested that Mary, mother of Jesus, may have done), and that rather than the heir of heaven Phaethon might just be another run-of-the-mill bastard. In despair at the taunting, Phaethon runs to his mother and begs quite reasonably, "Give me proof that my father is the sun god!"

Moved by her son's torment—and perhaps by fury at her tarnished reputation—Clymene swears to him that she hasn't lied, and suggests that he visit the sun god and demand evidence of his parentage. Assured by his mother that his father rises each morning not far from where they live, Phaethon optimistically heads eastward, crossing frontiers to India, the land nearest to the fires of heaven.

The sun god's palace is resplendent with bronze, silver, and gold; Phaethon stops at a distance because at first his eyes can't bear the radiance. The scene he eventually sees portends the themes that we will explore during the rest of our natural and symbolic day. The deity who may be Phaethon's father sits on his throne, clothed in sunrise crimson. Ranked on each side of him are the Hours, Days,

Months, Years, Centuries—evenly spaced apart. Each unit of time is born from and serves the sun or moon. In this story, as in others that we will explore, the hours and days blur with the seasons and the ages of humanity. Spring wears a crown of flowers; naked summer carries sheaves of grain; autumn's feet are stained with trodden grapes; winter has hoary locks.

Apollo asks the trembling youth, "What do you seek from me," and adds the yearned-for endearment, "my son?" Phaethon's worries tumble out. Apollo assures him that his mother is guilty of no indiscretion, that he is indeed the off-spring of the sun. And then Apollo makes one of those stupid mistakes that keeps the plot going in so much of mythology, opera, and movie thrillers. In a surge of paternal guilt, he promises his son any favor he may ask. The qualifier "reason-able" might have saved the day, but Apollo imposes no limitations—and almost before he can finish speaking, Phaethon asks to drive the chariot of the sun for one day.

Rather than withdrawing his irresponsible offer, Apollo protests that his son is mortal and that, other than himself, not even the immortal gods have enough power to control the cycle of the day. He warns that the morning road is steep and the view below dizzying, that the sky is continually awhirl with dizzying motion, that the driver must run the gauntlet of the zodiac—Lion, Scorpion, Archer. He confesses that he can barely control the horses himself. And then Apollo makes a beautiful point. Declaring that because what Phaethon truly wants is evidence of parentage, Apollo says, "Surely I give you proof of that by worrying about you as I do. My fear for you proves me your father."

It would be irreverent to suggest that this myth, which has been popular in the arts for millennia, boils down to a son borrowing and wrecking his father's car. Yet among the spendthrift asides that enrich Byron's *Don Juan* is his remark that had such delightful turnpike roads existed in Phaethon's time as were being built in his own, Apollo would have advised his son to fulfill the urge to drive by riding the York mail coach.

But Phaethon rejects his father's belated prudence and insists upon the favor he was promised. He climbs aboard the gleaming chariot. Aurora, goddess of the dawn, opens the crimson portals. The moon's thin crescent sinks from sight. Down below on earth mountaintops are reddening. Apollo commands the Hours to harness the winged, fire-spitting horses. Then he provides last-minute driving advice: to stay in the middle of the road and follow the wheel tracks that will somehow be visible in the air itself. He even insists that Phaethon wear sunblock; he rubs his son's face with holy balm. Phaethon takes the reins in hand. The barriers are removed and the eager horses launch into their route. Their hooves cleave the clouds. When Phaethon reaches the . . .

But why tell all of his story at once? We can return to it as each phase becomes relevant. Myths, like the rest of litera-ture—like every repetition of the ancient day—take place in the present tense. Phaethon's story takes all day to unfold, and we have all day to follow him.

Chariot

Apparently every early society personified the cycles of nature. If the sun traveled across the sky from east to west, its watchers imagined that it must be carried by some creature traveling in a manner similar to their own capabilities. Some North American and Australian myths envision the sun walking. Not surprisingly, the Egyptians, whose lives were ruled by the flooding Nile as much as by the solar god Ra, depicted the sun steering a boat from dusk to dawn, down a river on the dark underside of the world.

In many cultures, however, the sun performs his daily rounds upon an animal or seated in a vehicle pulled by an animal. Because most European cultures invented wheeled chariots, frequently we find recurring imagery about the chariot of the sun. Myths from the Alps to the Himalayas portray celestial deities driving the solar chariot daily across the arc of the sky. The Nationalmuseet in Copenhagen houses a gorgeous example of Bronze Age iconography, rescued from the Trundholm bog in Zealand: a cast bronze horse drawing a bronze wagon that carries upright the disk of the sun. The solar disk itself is formed of joined convex halves covered in wrought sheet gold that still vividly impersonates the sun's radiance.

Both the disk and its conveyance are worth a closer look. Innumerable cultures have portrayed a solar disk, either pictographically or in wood, stone, iron, bronze, or some other material. Because the sun flies across the sky, its disk was winged when representing Ra in Egypt or Ahura Mazda in

Iran or Shamash in Sumer. Almost three thousand years ago in Nimrud, a sculptor carved in basalt a scene showing the winged sun watching overhead as a man kneels before a potentate. But there were many other variations. Krishna's disk was flaming and Vishnu's rayed and spinning—the latter sometimes in the form of a discus, the *cakra,* a spoked and razor-sharp sun-disk that he flings at his enemies.

The chariot adds another element to solar imagery. The invention of the wheel was a milestone in both technology and metaphor. In a conceptual leap as lost in antiquity as the first wheel itself, the new round technology began to represent both motion and the sun. A spinning disk with a hole through the middle came to symbolize the rotation of the celestial sphere around its axis. By the time wheels were spoked, joining the nave and perimeter with radial supports, more than ever they reminded people of the sun—or at least of already common graphic representations of it. This is the same kind of pictographic cross-referencing that adopted the round eyeball into solar myths, as in ancient Chinese, Japanese, and Egyptian images of one eye representing the moon and the other the sun, or the Greek idea that the eye represented Apollo as viewer of the heavens. In some ancient European midsummer festivals, celebrants constructed wheels solely to set them afire before rolling them downhill into a river. An actual wheel wasn't even always required; a torch could be whirled in circles until its light blurred into a wheel shape that actually existed only in that pliant medium, memory.

If the sun is a chariot or is transported by one, it must be powered by some kind of motive force. These stories evolved many thousands of years before a vehicle could propel itself; therefore animals pulled the sun's car. Transportation at the time was capricious. It had to take into account the unpredictability of livestock that could become hungry and tired or could be distracted from their task. Moreover, someone must harness and drive these creatures, a requirement that throws into this daily cycle the universal X-factor in mythology, as in the rest of life: the unpredictability of human loyalty, motivation, and diligence. This kind of myth, incorporating quotidian concerns such as transportation variables and their consequences into a story about celestial rhythms, reminds us yet again how we make gods in our own image. Reconstructing tales backward from such clues, reverse-engineering folklore, we glimpse how the realities of everyday life shaped the evolution of each myth, and how every new element (wheel, chariot, horse, driver) contributed traits that turned the story in a new direction.

Inevitably, in some myths the chariot of the sun goes astray, with disastrous consequences. Anthropologists and folklorists have recorded innumerable cautionary tales about the problems resulting when nature's cycles stumble. Such fables recur throughout the world. Accounts of irresponsible deities may be rooted in the same emotional urges that inspire detective novels about murders in the vicarage and films about gigantic mutant ants. This kind of story permits us to surrender to the decadent thrill of chaos while trusting that order—good old steadfast order—will be comfortingly re-established.

In his *Poetics* Aristotle describes *catharsis* (from the Greek meaning "to purify") as a purging of the emotions through vicarious involvement in a narrative that inspires fear and pity. It is axiomatic nowadays that after Hiroshima we dealt with worries of nuclear radiation partially through monsters such as the giant ants in *Them!* and the fire-breathing dinosaur in *Godzilla*. Some commentators argue that the plague of vampire stories in the 1980s and '90s may have been a cathartic response to the blood-wasting disease of AIDS. Perhaps it isn't surprising that ancient mythology dwelt upon the loss and return of essential features of everyday life—including the sun itself, whose reliability was brought into question by clouds, eclipses, winter, and even nighttime.

Dance

The Day of the Sun

By midmorning on weekdays most people are either at their jobs or at school. Rush hour is over. The light is more white, less golden, and brighter. The air is becoming noticeably warmer, except on the most severe winter days. If there was fog it is usually lifting or gone by now. Mental fog is also lifting, bringing the time of day when we are likely to be doing our best work: post-coffee, pre-lunch, not yet worn down by the day's routines. The sun is high enough to be less apparent to us, more of an overhead presence above the visor of our eyebrows, the romantic sidelighting of early morning giving way to less dramatic overhead brightness. This diffuse all-penetrating light is what we mean when we say "the light of day" or "full daylight." This is rational light—Apollo's light.

Apollo's insistent fire shows up in even the very names of the days, reflecting our preoccupation with the sky and our conviction that patterns and correspondences are so important throughout the cosmos that we must conjure them when they aren't apparent. Some scholars even conjecture that the week has seven days to correspond with the seven celestial objects visible to the unaided ancient eye: five planets plus the sun and moon.

It is ironic that the English name for the Christian sabbath, which honors a god who jealously prohibits worship of other deities, also pays tribute to the much older sun god. Our word "Sunday" is at least a thousand years old. In Latin, five names for the days of the week were based upon specific gods, such as *dies Martis,* the "Day of Mars." With the Roman god of war translated into the Germanic god of war, Tiw (now usually called Tyr in English), *dies Martis* became *Tiwesdæg* and in time Tuesday. The same pattern survives in the other days of the week. Weekend days, however, were named after the dominant celestial bodies rather than individual mythological characters: *solis dies* (sun's day) and *lunae dies* (moon's day). *Solis dies* was in turn a translation of the Greek *hemera heliou*—all meaning Day of the Sun. When the Roman influence reached far enough north, the Germanic peoples of Europe translated these names instead of borrowing them outright. The Latin *solis dies* became in Old English *sunnandæg* and *lunae dies* became *monandæg,* eventually evolving into Sunday and Monday, and remembering the sky's lights even on cloudy days.

A strict law bids us dance.

—Kwakiutl teaching

The Law of Dance

The Just-So conjectures of mythology intuited the deep physical connections between light and our bodies, our factory-issue biases manufacturing chemical signals to keep us on track day after day. Myths and folklore about the world beyond our bodies demonstrate other physical and emotional aspects of daily rhythms, and they also express our awareness of a key aspect of nature—its restlessness. But their ancient authors could not have envisioned the astronomical realities behind the restlessness. Why does daytime alternate with nighttime? Why does the sun arc across the sky? Why does the moon wax and wane? Because, a Zen astronomer might explain, each phenomenon participates in the cosmic dance.

※

Everything moves.

The Book of Common Prayer pointedly sorts humanity into the quick and the dead for a reason: because we think of movement as defining life. Obviously a ballerina's pas seul echoes a marsh hawk's elegant banking maneuvers and the choreographed veer of a school of angelfish. In this fidgety universe of ours, however, even the inanimate seems charged with energy. Wind bursts out in tantrums; rivers seek the ocean; the heavens fling hail. In the early nineteenth century, when the Scottish botanist Robert Brown first observed through a microscope the haphazard motion of suspended inanimate particles now called Brownian movement, it

seemed to be the deliberate swimming action of tiny creatures. Then Einstein demonstrated in his busy year of 1905
that this microscopic jiggle is actually caused by the impact of
randomly dancing molecules. It was the first proof of the ven-
erable Greek notion that matter is composed of impossibly
tiny particles—and it also showed that those particles simply
cannot be still. Electron microscopy delved more deeply, vio-
lating the privacy of the molecule itself, bringing us visions of
a miniature cosmos deep inside all of matter. And those
uncertain electrons whirling *somewhere* in Zen emptiness
have made us realize that the very stones beneath our feet
are as loosely woven as lace.

As we climb up to the planetary level, we find that it is
motion—the rotation of Earth on its axis—that keeps our
planet almost round. Eons of a daily turn on the potter's
wheel molded our psyches even as it fashioned our planet
into its present shape—an oblate spheroid, flattened at the
poles and gaining weight at the equator as a result of its primordial spin on its axis: a middle-aged planet bulking up
around the waist. Furthermore, the tilted Earth heats
unevenly as it rotates, sending masses of warmer and cooler
air migrating across the globe and generating weather
changes, some of which we will examine this afternoon.
From far away the moon jostles the tides and they slosh
against the lip of their basin on Earth. Our planet and its
satellite are dance partners because of gravity, and both in
turn pirouette to the even stronger laws of the parental sun.
Nor is the sun free of such obligations; it rotates on its own
axis, the cycle of which can be measured by tracking

sunspots' movement across its face. Farther out, the planets lope in stretched elliptical orbits around the sun. Move upward and outward another step in this scenario, and you find our modest little solar system gliding along the edge of the dance floor with the rest of the galaxy.

Everything moves.

※

Yet it took human beings a surprisingly long time to figure out that no part of the cosmos is standing still, not even the earth beneath our feet. Ancient skygazers assumed that Earth is stationary. Even after the realization that the world is round and somehow hangs in space, there still seems to be plenty of evidence that it floats firmly in the center of the universe. Does not the sun circle us to light our path?

In the second century the Greek astronomer and geographer Ptolemy (Claudius Ptolemaeus) presented a beautiful mathematical model that explained how an Earth-centered cosmos worked. His complex system of circles within circles, of cycles and epicycles, employed centuries of Babylonian observations, explaining—and predicting with a fair amount of success—the motions of the planets. Remember that at the time Earth was not considered a planet. The etymology of this word tells us why: most of the night sky was thought to be so reliable that less well-behaved lights were dubbed *planetes,* "wanderers," because they didn't traverse the sky in the same path every night as did the "fixed" stars. Ptolemy's work was adopted into Christian cosmology, because it seemed to go a long way toward reconciling revelation- and observation-based worldviews. But it also helped establish a notion that would eventually undermine theological cosmology: an integrated secular universe that did not require divine micro-management.

Ptolemy's thought was hugely influential for more than a millennium. It greatly improved marine navigation, although its built-in errors still caused problems at sea, resulting in loss of lives and cargoes. In the Middle Ages his *Almagest* was held in almost religious esteem by both Christians and Muslims, even though his *Geography* and world map had long since passed their expiration date. The only problem was that Ptolemy was mistaken. Irrefutable evidence against him—and a more accurate alternative—didn't appear until 1543.

For two reasons, this year is often cited as the beginning of modern science. First, a Brussels-born anatomist named Andreas Vesalius published *De Humani Corporis Fabrica* (On the Structure of the Human Body). This insightful and beautifully illustrated volume corrected a number of errors in the work of the second-century Greek anatomist Galen, who over the centuries had been practically beatified by the church as the Aristotle of the human body. The second crucial event in the history of science that occurred during 1543 took place only one week after the appearance of Vesalius's magnum opus: the publication of an insurgent masterpiece called *De Revolutionibus Orbium Coelestium* (On the Revolutions of the Heavenly Bodies). Its author, Niklas Coppernigk, whom history remembers by the Latinized form Nicolaus Copernicus, was born in what is now Poland and spent much of his adult life as canon of the cathedral in Frauenburg. When he wasn't physicking the poor or fulfilling other clerical duties, this bourgeois gentleman spent his time on mathematics and astronomy, cleaner and less rank pastimes than Vesalius's. In 1543 he was seventy. He had been working on his astronomical studies for many years—he had shown friends a summary twenty-nine years before, the year that Vesalius was born—and had completed *De Revolutionibus* in 1530. Apparently he delayed thirteen years in publicizing his findings because he feared both academic laughter and reprisal from the church.

Like Vesalius, and like many scientists ever since, Copernicus focused his work around a refutation of an earlier scholar. He argued that Ptolemy's system was incorrect, that his evidence did not support a geocentric theory, that some of his points introduced unnecessary complications and *still* failed to prove his case. "Hence a system of this sort," he wrote in his historical summary of the issues, "seemed neither sufficiently absolute nor sufficiently pleasing to the mind. Having become aware of these defects, I often considered whether there could perhaps be found a more reasonable arrangement of circles, from which every apparent inequality would be derived and in which everything would move uniformly about its proper center, as the rule of absolute motion requires."

Note that Copernicus was committed to absolutes and ideals. And, like natural philosophers from Euclid to Stephen Hawking, he spoke of beauty and simplicity as much as of the accuracy of his model. As usual, the caricature of scientists as robotic fact-machines deflates in this scenario; Copernicus was fueled as much by Renaissance aesthetics as by a desire to rescue navigation and reform calendars. Scholars such as Owen Gingerich argue that in reality the Copernican model wasn't much simpler than the Ptolemaic, and not dramatically more accurate, but that for the first time the heliocentric view was a serious competitor. Finally it seemed at least as viable as the geocentric.

Where Ptolemy began with Egyptian computations, Copernicus turned to Greek writers who had suggested two millennia earlier that Earth actually orbited the sun rather than the other way around—Philolaus, who lived in the fifth century BCE, and Aristarchus of Samos. Aristarchus had been (and still is) too often ignored by nonspecialist historians, partially because he didn't derive from his theory any better calculations of planetary positions than were already available from geocentric models. What Copernicus did not realize—what he could not envision without knowing the role of gravity and momentum and inertia in celestial mechanics—was by what law the planets could dance around each other with nothing to hold them in place.

This unavoidable ignorance is one reason why he retained in his system the hoary but oddly beautiful fabrication of the celestial spheres—concentric invisible spheres nested each within a larger, beginning with the moon and working outward to the planets and a single background sphere of stars. Actually, the translation given above for Copernicus's masterwork—the usual English name for it nowadays—isn't quite accurate. *Orbium Coelestium* translates as "Heavenly Spheres," not merely "Heavenly Bodies." Soon Tycho Brahe in Denmark would undermine the notion of the perfect unchanging sphere of stars with *De Stella Nova*, his interpretation of the "new" stars that suddenly burned in the night sky, challenging assumptions about stars' fixity and permanence.

It turned out that the church didn't get around to attacking Copernicus's work until long after he had gone to his reward. The pope was busy elsewhere. In England Henry VIII had appointed himself secular pope, and in Germany Martin Luther was denouncing corruptions infesting the Catholic Church, especially the sale of indulgences (remission of punishment for sin). Trouble was brewing everywhere. Who had time to slap down astronomers? The censors' oversight bought time for the ideas of Copernicus to take root. They quickly did so across Europe, among astronomers, physicists, and mathematicians; it was 1616 before the church placed *De Revolutionibus Orbium Coelestium* on its list of forbidden texts, that elite must-read index of scientific and literary works. But by this time neither Copernicus nor history cared. The church only did so then because the telescopic discoveries of Galileo created a much wider awareness of Copernicus and his blasphemous views. Galileo had been busily watching sunrise on the moon and sketching further evidence of the dancing universe. We will peer through his telescope this evening, as the moon rises over our own night.

Not Invisible

Although the light at this time of day isn't as intense and varied as at dawn or sunset, it has its unique characteristics, and attending to them can add a quiet pleasure to our late morning. While doing so, let us turn to the private musings of an insatiably curious man who, more than half a millennium ago, paid affectionate close attention to the world around him. In his notebooks, Leonardo da Vinci distinguished between three kinds of perspective. The first is the phenomenon that most of us think of when we see the word *perspective:* the apparent convergence of parallel lines as they recede into the distance. This *linear perspective* is demonstrated to us throughout the day by streets, train tracks, sides of buildings, even the view from one end of your dining table to the other. Think how odd this optical illusion would look to us had we not become inured to it since birth; it is said to be one of the most shocking visions for people who, blind or almost blind from birth, have had their sight restored through surgery. It is the kind of perspective in painting that depends upon drawing ability rather than color mixing, but fortunately there are mechanical aids to assist us.

Naturally Leonardo didn't stop with the most obvious optical illusion, and it is his other two that particularly interest us today. The apparent haziness of objects at a distance is the result of *aerial perspective,* and no one has described it more lucidly than Leonardo himself: "The impact of the appearance and of the substance of things diminishes with every successive degree of remoteness; that is, the farther the object is from the eye, the less will its appearance be able to penetrate the air." This kind of perspective results from what physicists call *absorption,* which is simply the reduction of the amount of light reaching our eyes, making distant vistas both less distinct and dimmer in color.

Such fading away into the distance isn't surprising, considering the obstacle course that light has to run on its way to our eyes. The atmosphere transports an almost unimaginable weight of floating objects: particles of dust and pollution, infinitesimal seeds, near-microscopic animals, wisps of bird feather, human dandruff, dried beetle excrement—the list is endless. We've all observed the impressive population of dandelion or milkweed fluff that can hitch a ride on an autumn breeze. In the 1930s a U.S. entomologist made almost fifteen hundred flights over Louisiana, scooping up insects into screens that stretched between his biplane's wings. At altitudes from just above the ground to 15,000 feet, he collected over 30,000 insects that he later painstakingly sorted into more than 700 species. Elsewhere, other researchers were puzzled as to how a jumping spider could survive in the unkind crannies atop Mount Everest—the highest-altitude habitat on our planet—only to discover, upon further exploration, that the spider was having care packages delivered every day, as countless minuscule insects were airlifted from the valleys below. After meticulous analysis of air samples at every height and over all kinds of terrain, scientists estimated that a single cubic mile of air over the more fertile temperate areas on Earth could contain as many as 25 million organisms. Light must jostle between this airborne multitude to reach our eyes, and it isn't surprising that longer and shorter wavelengths arrive in different ways.

Ignoring these rules of physics in art can create striking effects. Certain Renaissance paintings, such as Jan van Eyck's *Madonna of Chancellor Rolin* in the Louvre, present their symbol-saturated portraits and settings illuminated by a diffuse indirect light, often one in which atmospheric perspective is trumped by artistic license: all objects near and far are equally detailed and vivid. Unless the maker of the tour de force provides clues, such as an open book of hours, it is often difficult to determine the time of day. The source of lighting is not the sun at a particular location in the sky but rather the divine all-seeing light of God's wisdom.

Just as visible as atmospheric perspective, although we seem to consciously notice it less often, is *color perspective*. "The greater the depth of the transparent layer which lies between the eye and the object," wrote our tireless noticer, back in the Renaissance, "the more the color of the object will be modified by the color of the intervening transparent layer." For example, the graduated blues of increasingly distant peaks in the Smoky Mountains occur partially because the intervening layer of air isn't truly transparent—as in the phenomenon of aerial perspective—and partially because, although the eye detects relatively little light from distant green foliage, it also collects the light scattered in the air between it and the trees, and this light is mostly blue. The physicist and excellent science writer Hans Christian von Baeyer has pointed out that it isn't as if no human being ever noticed that faraway mountains appear blue; they are portrayed this way in Roman wall paintings. The effect seems to disappear, however, until its gorgeous rebirth in the background of the Flemish landscapes of the Renaissance. "The air is not invisible," wrote von Baeyer. "We see it all the time."

The Constellations Flee

As Phaethon grasps the reins in his young and inexperienced hands, Apollo's winged steeds—Aethon and Pyrois, Eous and Phlegon—paw the air, eager to be off on their daily journey across the sky. The chariot rides lightly behind them, swaying like a ship without ballast. Unaccustomed to Phaethon's lesser weight and unnerved by his trembling guidance, the horses lunge and snort. The reins are now limp, now taut, and the boy's terror channels through them and into the horses. The animals glance back over their shoulders, the rolling whites of their eyes brilliant among the red clouds of sunrise. With a nauseating rush of fear, Phaethon realizes his mistake in the very instant that he knows he cannot turn back. He can't turn in any direction. Already the horses are straying from their course. Phaethon can foretell that wherever the chariot goes he will be rider, not driver.

Frightened by this betrayal of the sacred daily routine, the constellations flee the fire of the wayward sun. Scorched, astonished, the Bears, Ursa mother and child, try to plunge into the cold northern ocean, but they are tethered to the pole star and, circle it though they may, they cannot escape. The great Serpent, usually dozing lethargic near the pole, feels the unaccustomed warmth and stretches, coils, uncoils; the fire feeds his poison and his latent rage. Even Boötes, the oxcart driver, tries to run across the sky, pulling at the heavy cart to escape.

Queasy, horrified, Phaethon forces himself to gaze down at the land far below. Under their watchful constellations, which daylight has made invisible, farmers—who started work at dawn to avoid the midday heat—halt their oxen and shade their eyes with callused hands. They peer upward with the kind of fear they usually reserve for an eclipse. From Phaethon's high vantage, one farmer stands at the end of an unfinished furrow like a pen pausing midsentence. City dwellers, strewn like chess pieces across the square stone blocks of streets, gaze at the flaming sky. Phaethon knows that with the arrogance of youth he has betrayed Apollo, Clymene, himself, the chariot's loyal horses—but worst of all, he realizes as the chariot careens across the sky, worst of all he has betrayed the trusting earth below.

Mere Fire

What is the sun?

As the astronomical body in question approaches zenith over our heads, let us try to be objective and scientific in answering this important question. What is the *real* sun, the sun that warms those creatures who lack symbols, who don't personify natural processes as runaway chariots? Never mind the solar god who jealously melts the wax in Icarus's wings to punish his hubris. We're talking about the nearby star that hides behind Earth during nighttime on your side of the planet (wherever you may be as you read this sentence), the ball of fire that blazes in the darkness of space, blithely unaware of the millennia of hosanna and entreaty that we have draped around its royal shoulders.

In an early poem Richard Wilbur wonders what Earth would be like without human beings to interpret its phenomena, with "a stone look on the stone's face," with the sun reduced to "mere fire." Do the oriole and manatee perceive the sun more directly because they lack a story for it? It seems unlikely. Granted, before us dawn meant not the promise of renewal, not infancy, not the antonym of twilight—just the return of daylight and the ability to see more clearly. Sunset was merely the departure of light, not the autumn of the day or a prelude to symbolic death.

Removing the mythology—as much as we can even begin to accomplish such a task—won't reduce the sun to a fire without story. In our attempt to see the sun as it truly is, we can concentrate on what science has learned of its long and flamboyant history. And what could be more poetic than the bare facts of the cosmos?

✺

If there is nothing new under the sun, at least the sun itself is always new, always re-creating itself out of its own inexhaustible fire. Every moment our neighborhood star squanders a caliph's wealth of energy, flinging it in all directions. Nuclear fusion supplies it; the sun is a roiling mass of raw elements. Like other stars in its class, it is about 71 percent hydrogen and 27 percent helium, with the remaining smidgen comprising carbon, nitrogen, oxygen, and other elements. At its core the sun burns at a temperature of 15 million degrees Celsius (27 million degrees Fahrenheit). On most of its surface the sun's temperature is a comparatively frosty 5,500 degrees Celsius (9,900°F), plummeting as low as 3,700°C (6,700°F) in the dark centers of sunspots.

Such a hellish internal temperature, combined with a density that we can mathematically estimate but not truly envision, strips the electrons from hydrogen atoms. This process leads to the nuclear fusion that powers the universe: naked hydrogen nuclei colliding and fusing into helium atoms. Via the terrifying equations that Einstein found hidden like the Ark of the Covenant, the conversion of four hydrogen nuclei into a single helium nucleus destroys a small amount of mass but thereby produces a vast amount of energy. It seems as if there could never be enough of it, however, because the sun demands as fuel 4.3 million metric tons per second. Clearly the sun's fire isn't truly inexhaustible. Astronomers prophesy that its supply of hydrogen will last for a few billion years, but what will happen when it runs out of energy? We will save for this evening the fanciful scenarios about the sun's demise; it seems appropriate to consider the last sunset during the daily one.

As closely as one can measure a burning ball of gas, the sun has a diameter of 900,000 miles. Sunspots alone—and they're just the cool areas—can extend as far as 60,000 miles, wider than seven earths, a quarter of the distance from us to the moon. And yet the supreme deity in our sky is a middling star, one of interchangeable billions so common as to be vulgar. The sun's nearest neighbor, Proxima Centauri, is so distant it seems irrelevant by comparison—2.2 light-years, 275 times farther away than the sun. Remember that this means that when you look at it you are seeing how it looked two years and two months ago, more than half the duration of a college education. Such distance reduces it to a feeble one-forty-billionth of the sun's luminosity, scarcely enough for a few emissary photons to reach our eyes. And this is the *closest* star. Ours is a lonely neighborhood.

Because we experience the sun as the dominant presence in our local cosmos, it is humbling to realize that not a tithe of its affluence reaches the surface of our planet. Interestingly, despite the sun's inner turmoil, the intensity of all wavelengths of radiant energy that arrives at the Earth's outer atmosphere varies by only 0.1 percent. A physicist will tell you that Spaceship Earth receives a solar energy allocation of 430 BTU/ft^2/hr, which comes to 1.37 kw/m^2. To translate the abbreviations into a measure of energy with which most of us are more familiar, our atmosphere receives roughly two calories per square centimeter per minute.

Those elementary school experiments that portray Earth as a pea fifty yards from a basketball Sun remind us that our star radiates its energy in all directions equally, and that Earth is not the sole recipient of its distant light. To imagine that the sun was made for our convenience is like envisioning millions of dollars being showered at random for the sole advantage of a distant cup that catches a penny. "We are receiving our portion of the infinite," wrote Thoreau, and it begins each morning with the rising sun. A tiny essential splash of solar energy is captured every day by the single flower in this barren field: the Earth in space.

Although hydrogen and helium do exist and Apollo and his chariot do not, it is beginning to look as if mythology has not overstated the sun's importance—neither its generosity nor its tantrums. The scientific evidence indicates that the sun's royal position in the mythological hierarchy makes perfect sense. Ours may be a negligible star tucked away in the galactic boondocks, but until the extinction of *Homo sapiens* it will never be mere fire.

High Noon

I did so love my body
And the wide noon! . . .
I was a shape that made a shadow
And my flesh glowed in the sunlight.

—Don Marquis, "The Ghost Speaks"

Shadow Show

History, mythology, the sciences, the arts—all agree that unexpected significance lurks in the overlooked phenomena of everyday life. Consider this fact of existence: when light strikes an object, the object will cast a shadow. What could be more obvious? Our ancient brains were programmed with such information long before we evolved into a species that would dub itself wise. Except very early or late in the day, we don't notice such seemingly trivial occurrences—and we don't need to.

Whether unconsciously or deliberately incorporated into your aesthetic awareness, however, your perception of shadows is an inescapable part of your experience of the day. Representational art would be crippled without shadows and shading. One of the classic scientific insights, formulated over two thousand years ago, involved a noontime shadow, as did a discovery about the nature of timekeeping. We will get to those stories soon, but we will begin our look at shading and shadows modestly, with a little boy named Milo.

Milo is the hero of Norton Juster's splendidly imaginative children's book *The Phantom Tollbooth*. One day the bored child receives from a mysterious benefactor a child-size car and matching tollbooth. Immediately he seats himself behind the wheel and drives the car through the magical tollbooth into worlds of adventure populated with characters such as the Spelling Bee, the Humbug, King Azaz the Unabridged, and the world's shortest giant and tallest dwarf—who happen to be the same man.

During his journey, Milo witnesses the concert of the day's light. The conductor, Chroma the Great, directs the silent thousand-piece orchestra whose daily performances create the ever-changing colors of the day. Finally the newly inquisitive Milo asks Chroma what the world would be like if the players stopped performing. The conductor raises both hands high. At this gesture, all the instruments that have been playing cease their music—and all color in the world vanishes. The world appears as a black outline drawing waiting to be filled in. "You see what a dull place the world would be without color?" asks Chroma.

Actually, of course, Juster's description of the halted symphony makes a conceptual leap beyond colorlessness. Loss of color does not mean a loss of graded hues and textures. A world without color would not suddenly look like a coloring book; it would merely look like a beautifully detailed black-and-white photograph—or, because it would not stop moving, like an old movie. Outline drawings express another theoretical leap entirely. Line, itself a convention, is the representation of abutting planes and shadows, or the apparent overlap between objects of varying distance from the viewer. A colorless world would indeed be dull, but it would be even duller without gradations of tone.

※

Throughout history, despite allegorical rationales that sneaked a nude past religious censors, despite the expectations of royal patrons, despite conventions of genre or era, the subject of many paintings has simply been light itself. One of the most exquisite visual pleasures in life results from the wealth of artistic responses to the play of light on the divers textures of our gloriously material world. The convincing roundness of painted objects in representational art, all of their illusory three-dimensionality, results from the artist's re-creation of light and shadow.

So automatic is our response to this deception that we take for granted the process of creating and perceiving it. If you stop to look closely at your computer screen, you will find that we even use our built-in awareness of the rules of light and shadow to trick the brain into finding new technology reassuringly familiar. Many Web sites and e-mail systems portray a link to another page as a "button," complete with shadow and a fake highlight from some imaginary overhead lighting source; usually, when a cursor clicks on it, it mimics being pushed by changing the location of the highlight and shadow.

If shading is the essence of convincing representation even on your computer screen, think how much more important it is in artwork. Representational painters spend most of their time impersonating the sun's caress. Think of the subtle modeling of skin tones in Roger van der Weyden's symbolically charged Renaissance gardens, or the haunting photorealistic streets of Richard Estes. Throughout his three-dimensional toy universes, Canaletto's egalitarian sunlight delineates architraves as gently as it implies the shadows of pigeons. Even distant figures cast believable shadows across convincing streets, a sight that without our realizing it persuades and lures the viewer. No example makes this point better than its opposite—painters such as the American kitsch franchise Thomas Kinkade, purveyor of scenes featuring artificially sweetened illumination without source or rationale.

A glance back into history demonstrates this point. Completely different kinds of attention to light and shadow separate the black-figure pantomimes on ancient Greek urns from, say, the three-dimensional-looking shading in Michelangelo's red-crayon studies for the Libyan Sibyl. Between these examples there gapes a chasm of slowly evolving consensus about how to perceive and represent nature, centuries during which artists refined representation toward an ever more convincing illusion. Part of what made comic strips such as Bill Watterson's *Calvin and Hobbes* and Berke

Breathed's *Bloom County* so visually impressive was their dramatic use of light and shadow, even without the subtleties of shading. Comic books attained a new level of sophistication and appeal when printing techniques—and the arrival of ever more stylish artists—resulted in three-dimensional works such as Alex Ross's luscious 1996 graphic novel *Kingdom Come*.

The shadows beneath the figures in paintings are worth our attention. Some painters emphasize shadows within a scene—separate from the modeling of bodily contours—and some almost ignore them. For most of his early *Saturday Evening Post* covers, Norman Rockwell posed characters and perhaps an article of furniture against a white background. One feature of the accidental unreality of his later paintings is that the human figures appear to have been painted from models posed indoors, after which Rockwell inserted them into outdoor scenes of which they are clearly not a part. His autobiography confirms this suspicion. This disjuncture is especially true of his more socially aware paintings, such as the one of white children watching black children move into their neighborhood, or the famous 1964 *Look* illustration in which a black girl walks into a new school surrounded by white U.S. marshals. All of these figures may as well be standing or walking on air, they seem to have so little connection to their surroundings—despite the pollock of thrown tomato on the wall behind the lonely girl. It is no coincidence that Rockwell's few paintings to approach the level of art, such as *Shuffleton's Barber Shop*, employ a more holistic approach,

integrating figure and setting primarily through the relationships between shadows.

Children paint shadows as if they were black, as do many
adults who don't normally think in terms of translating a
scene into a picture. In a line drawing shadows often wind
up approximated with cross-hatching, masses of pointillist
dots, or a carefully placed mass of pure black. Part of the revolution in painting launched by the Impressionists was a
new awareness among such painters as Auguste Renoir and
Claude Monet of color in shadows. Eugène Delacroix shook
up the stylized elegance of painters such as Jean-Auguste-
Dominique Ingres by returning from North Africa with a
new sense of color and shadow inspired by the stronger
southern light. Monet too had observed the harsh contrasts,
the complementary-toned shadows, of northern Africa. By
the time of Henri Matisse, a creative *rejection* of shading and
shadows appeared revolutionary. The relative lack of shading in Japanese woodblocks—the strong outlines and
emphasis on pattern—inspired Vincent Van Gogh. In the
famous picture of his room in the yellow house at Arles, Van
Gogh eschewed anchoring shadows, relying instead upon
outline and color.

The first problem for painters is that shadows don't stay
where the artist wants them. The sun arcs swiftly across the
sky. Plein air painters have always had to work at top speed,
unlike still-life artists who can position their subjects before
an obediently stationary lantern or electric light. Any Sunday
painter is familiar with the routine: to paint the simplest

landscape requires several visits to the scene, each at the same time of day and if possible in equivalent weather. Gradually a distillation of various days emerges on the canvas, as shades of paint mimic light and shadow to lend two dimensions the illusion of three.

※

Light may be the subject of much of painting, but it is the very medium of photography. It is unfortunate that the invention of photography was thought for a while to supersede realistic painting, because the two processes employ different methods for different goals. Naturally photographers and cinematographers also pay close attention to shadows and shading. Just think of the voluptuous peppers and nudes of Edward Weston, or the luminous portrait of a young Ellen Terry by the nineteenth-century pioneer Julia Margaret Cameron. Ansel Adams devised the Zone System to encourage photographers to "pre-visualize" the range of shades in a scene before attempting to capture it with a camera. Of course, the Zone System was aimed at black-and-white photographers; color shooters may choose to disregard subtleties of shading in exchange for beautiful colors, but most of the points apply to both. As every aspiring photographer—rather than weekend snapshooter—soon discovers, the tonal range available in photographic negatives and papers is exasperatingly finite, and a wise photographer determines camera settings and exposure times with the final image in mind. Even digital snapshots demonstrate that attempting to record too great a range of light and shadow, or too dramatic a contrast, will reduce detail and overall quality in a photograph. Perusal of Zone System manuals—oddly poetic technical handbooks—can rouse lackadaisical eyes to the subtle tonalities around us.

We don't have time in our busy day to explore this vast subject at length, but we must include one particular image from art history that unites painting, photography, technology, myth, and light itself. This morning we discussed medieval sun-worshippers whirling torches in the nighttime air to create the illusion of a wheel shape. Hundreds of years later, after the sun's demotion to a knowable furnace had secularized light imagery, Picasso drew figures in the air with a flashlight, creating ephemeral works of art preserved only by slow-exposure film—reminding us that the word *photography* means "drawing with light."

Hunt the Shadow

Naturally artists have not been the only ones to notice the changing shadows of the passing day. Long ago this natural cycle of daily change, like lunar phases and tides and seasons, proved reliable enough to encourage measurement and inspire the invention of a clever and durable chronometer. Many early societies invented water clocks and hourglasses, but no timekeeping device is more elegant or versatile than the one that employs everyday sunlight.

The sundial, most versions of which tell time by casting the shadow of an upright *gnomon* (from the Greek word for "indicator") across a graduated base, is an ancient instrument.

Pliny the Elder claimed in the first century that Anaximenes of Miletus, a pupil of Alexander the Great, discovered the "theory of shadows." Pliny wrote of Anaximenes, "He first put a sundial on show at Sparta, a device they call 'Hunt-the-Shadow.'" Unfortunately the gullible Pliny was once again taking hearsay for evidence, because the invention unquestionably pre-dates Anaximenes. The earliest known versions of this instrument lacked even the now familiar round dial; the Egyptian time stick from the tenth century BCE cast the shadow of a crossbar on a horizontal track that was marked with only five points.

For centuries, sundials were essential items in everyday life. Methods of determining the time of day by shadow started out simple and rough, apparently involved mostly in religious rituals. Even a twig stuck upright in the earth will provide a makeshift clock, and from there the variations are limitless. In a sense, even Stonehenge and other megalithic monuments such as the vertical columns at Machu Picchu were calendar-size sundials. Over time, sundials became ever more beautiful and elaborate. Many in the West were inscribed with classical quotations and cheerful Latin mottoes such as *Mors certa sed hora incerta* ("Death is certain; only the hour is uncertain").

Considering how often in this book we return to the influence of Earth's rotation on its tilted axis, it is worthwhile to note that the great conceptual leap in sundials occurred sometime before 300 BCE. Originally the gnomon on such clocks was vertical, aiming straight up to zenith, and told time by the length of its shadow. The Egyptian time stick, for this reason, had to be aimed eastward at dawn and westward in the afternoon. The brilliant innovation was to tilt the sundial's gnomon at an angle corresponding to its latitude and point it north, thereby measuring time by the angle of the gnomon's shadow rather than by its length.

During the seventeenth and eighteenth centuries, shops in Europe were producing mechanical clocks at an impressive rate. The manufacture of sundials kept up with them, though, because the clocks' mechanisms were still unreliable enough to need a frequent reality check from sunlight and shadow. No wealthy country home was complete without a horizontal sundial ornamenting the garden, and many churches had vertical sundials on an exterior wall. Even people who carried a pocket watch also carried a pocket sundial. They did so, that is, until the middle of the nineteenth century, by which time the dominant timekeeper in most towns was a clock maintained by the local telegraph office, which usually kept Greenwich time. (As we will see shortly, the telegraph had a powerful impact on many ways of thinking about time that concern us in our trip through the day.)

A shadow cast by the sun has drawbacks as a timekeeper, and the adoption of standardized time was impossible prior to the invention of almost instantaneous communication. Even with the Royal Observatory at Greenwich established as the prime meridian of the globe, every locality had its own time as the Earth turned toward the sun. For example, Bath, England, located almost two and a half degrees of longitude west of Greenwich, would reach noon almost ten minutes after the hour had chimed at the Royal Observatory.

Brighton, on the other hand, located very close to the prime meridian, would not experience the same delay. Obviously, with so many astronomical mysteries to contend with, sundials were not as simple as they appear in a country garden. Building an accurate one required elaborate calculations, because hours would be equally spaced on the dial only at the equator.

In his old age, Thomas Jefferson liked to distract himself from his chronic rheumatism by calculating the hour lines for a sundial. The latitude of his Poplar Forest hideaway in Virginia, which he built as a refuge when Monticello became too well known for his taste and began to attract admirers, was 37°22'26"N. His calculations were typically ambitious— every five minutes to within half a second of a degree. The Alderman Library at the University of Virginia contains in its archives Jefferson's own instructions for the precisely aligned installation of a sundial. When the architect Benjamin Latrobe was designing the Capitol building in Washington, he sent the model to Jefferson. The former president admired the capitals on Latrobe's design but thought that they needed some utilitarian purpose, so he turned one of them into a base for a spherical sundial. When Jefferson sent Latrobe a sketch of his adaptation, the architect was highly complimentary about the design's classical simplicity. He suggested that everyone could have a good sundial if they followed Jefferson's ideas. But new and ever more reliable mechanical timekeeping devices were being invented, and soon the sundial would be demoted to a quaint garden artifact and a fancy perch for songbirds.

Measure the Earth

Nothing could be less substantial than the shadow dogging our steps or stealing across our sundials, yet this phenomenon was the primary tool in a pioneering scientific calculation that took place more than two thousand years ago. The history of science is rich in fascinating tales of how someone employed an overlooked everyday occurrence to launch grand discoveries, but few scientists have accomplished more with less than Eratosthenes.

He was born in the third century BCE in Cyrene, which is now the city of Shahhat in Libya. In an era of rampant disease and raging wars, he managed to live eight decades. In 236 BCE, at the behest of the ruler Ptolemy Euergetes—who is also remembered for decreeing a leap day to round out the already 365-day Egyptian calendar—Eratosthenes became curator of earthly knowledge by taking over administration of the fabled library at Alexandria, capital of Ptolemaic Egypt. His mathematical innovations included a mechanical system for doubling cubes and what we now call the Sieve of Eratosthenes, a method for determining prime numbers. But he ranged well outside math. He correctly explained the annual flooding of the Nile as the result of rain falling into lakes in East Africa, and he created a map of the world showing lines of both latitude and longitude. Unfortunately his longitudinal demarcations tended to run through large cities rather than remain evenly spaced. During the next century, however, Hipparchus of Nicaea (now Iznik in Turkey) improved upon Eratosthenes's model by spacing the longitude lines equally, creating the system that grids every map and globe today—a beautiful example of how we overlay nature with contrived guidelines.

But most important for our purposes today, around 240 BCE Eratosthenes cleverly employed shadows to calculate the size of planet Earth—long, long before anyone actually circumnavigated the globe.

※

To understand the reasoning of Eratosthenes, let's begin by consulting any contemporary globe. (As a bonus, we will acquire a better grasp of the nature of seasons.) Immediately you will find Eratosthenes's legacy in the latitude and longitude lines. But two other lines, both parallel with the equator, seem even more arbitrary. At a curiously uneven 23.5 degrees north of the equator lies the Tropic of Cancer, an imaginary line that, as it circles the planet, runs roughly through Mexico City, Havana, Riyadh, and Guangzhou. At 23.5 degrees south of the equator, the Tropic of Capricorn runs near Rio de Janeiro, Pretoria, Alice Springs, and Nuku'alofa.

Why are these lines in such odd positions on the globe? It is no coincidence that the Tropics of Cancer and Capricorn are the same number of degrees from the equator as the Earth tilts on its axis away from the ecliptic, the plane in which it orbits the sun. The word *eclipse* reminds us that such astronomical alignments are possible only because the bodies involved lie within the ecliptic plane—so that, for example, Earth's moon can block the sun, or Earth can come between the sun and moon. Earth's axial tilt unites with its elliptical orbit to produce seasons, which are caused by the considerable variation in the angle and intensity of solar radiation warming the planet during a year.

Every six months the sun shines directly down on the equator. The resulting twelve hours of daylight and twelve of darkness provide the name for these two days—*equinox* ("equal night"). Precisely a quarter of a year after an equinox we reach a solstice. Just as the sun shines directly down on the equator on the equinoxes, so does it shine directly down on the Tropic of Cancer on the summer solstice. The official beginning of summer and winter (reversed in the Northern and Southern hemispheres, of course) falls on these days. The location of the Tropics was determined in response to the apparent motion of the sun as the Earth tilts first one way and then another. To our earthbound eyes the sun seems to rise from constellations in the southern sky for part of the year and from those in the northern six months later.

The word *solstice*—which derives from *solstitium,* meaning the "sun stands still"—refers to the moment when the sun, against the background of the constellations from which it appears to rise, seems to pause at its northernmost or southernmost point before starting back in the opposite direction. The night sky demonstrates these changes. In the Northern Hemisphere, for example, the constellation Scorpius dips partially below the horizon during summer and then climbs higher again. Few of us watch the sky enough to even begin to notice such subtleties. For us the change in the cycle shows up as daylight hours growing ever longer after the winter solstice, until they reach their longest on the summer solstice, at which point they start to decline again.

꙰

Less than one degree north of the Tropic of Cancer, as we measure now, and about five hundred miles south of Eratosthenes in his library at Alexandria, lay the city of Syene (now Aswan). Erastosthenes learned that at noon on the summer solstice the sun was directly overhead at Syene. Its reflection appeared in the water of narrow wells, and shadows of upright poles were almost nonexistent.

The scant historical sources provide few details, but the next series of events may have gone something like this. Late one summer morning, on the solstice, Eratosthenes went outdoors at Alexandria and placed a rod upright in the sand. Then he waited for noon, which at the time meant simply the moment of shortest shadows. He knew that only at this exact time on this particular day of the year would the sun at Syene be directly overhead. Consequently he wanted to measure by how much of an angle the sun missed being directly overhead at the same moment in Alexandria.

When Eratosthenes measured the rod's tiny shadow at noon, its direction and length revealed the slight angle by which the sun was off zenith: 7.2 degrees. This meant that, out of a circle of 360 degrees, the percentage of the circumference of the Earth spanned by the distance between Alexandria and Cyene was 0.02, or one-fiftieth. Therefore the circumference of the Earth must be about fifty times the

distance between the two cities, or roughly 25,000 miles. Naturally the circumference also told Eratosthenes the diameter
of the planet. As you probably recall, the ratio between the
circumference and diameter of a circle is always $3.14159\ldots$,
expressed as π. Knowing the circumference of Earth, Eratosthenes divided it by π, getting a figure impressively close to
the current official diameter of 7,926.41 miles.

Though unquestionably close, the numbers above are estimates. There is no question that Eratosthenes was a brilliant
man, or that his calculation was an impressive feat of geometry, but there *is* a question about the degree of accuracy of his
measurement. Historians seem unable to figure out precisely
the length of the Greek unit of measure, the *stadion,* that
Eratosthenes employed. The stadion varied. Apparently it
originated as a standard length of a plowed furrow, 600 *podos*
(from *pous,* "foot"). Like feet and yards until modern times,
the length of a podos varied, and in consequence so did the
length of a stadion. (Incidentally, *stadion* came to be applied
to a race covering this distance, and because races were usually run before seated crowds we now have the word
stadium.) Although by the time of Eratosthenes there were
official lengths for stadia, they varied from place to place.
Attic and Olympian stadia, for example, differed by about
forty-five feet.

Chet Raymo, the lyrical American astronomer and science
writer, argues that the most profound insight of Eratosthenes
was his vision, despite all the hilly obstacles wrinkling Earth's
surface, of a perfectly spherical planet whose ideal circumference could be calculated. In this measurement, says

Raymo, Eratosthenes united for the first time in history an idealized conceptual model, a quantitative observation, and a mathematical computation. "Eratosthenes' drawing of the Earth as a geometric circle," declares Raymo, "represents something as formidable as the plays of Sophocles, the history of Herodotus, or Athenian democracy: a way of abstract thinking that would eventually carry human imagination to the far-off galaxies."

How odd to think that Eratosthenes accomplished this conceptual milestone with equipment consisting of a stick, a shadow, and geometry. Even if he was as much as a couple of hundred miles off the actual measurement, his was still a splendid achievement, and a brilliant use of such down-to-earth tools. And with this story—with the image of a clever, hardworking man sweating under the midday sun, measuring a fleeting shadow—our journey reaches the official middle of the day.

The Equation of Time

Naturally human beings did not stop measuring shadows when Eratosthenes died. More than two millennia later, on the other side of the world, a child sat on a rooftop monitoring the sun and learned about the difference between clocks and reality.

Edward Emerson Barnard grew up to become one of the most important astronomers of the late nineteenth and early twentieth centuries. He revolutionized astronomical photography and contributed to our knowledge of everything from planets to globular clusters. He discovered the fifth satellite of Jupiter and more comets than any other person in history. But when he was born into the slums of Nashville, Tennessee, just before the Civil War, his prospects were not encouraging. His father died before he was born. As a young boy he could hear the cannons of war in the distance; soon Nashville was occupied by Union forces and the formerly bustling city sank into economic depression.

In 1866, when he was not quite nine years old, young Edward began his first job. He worked for John H. Van Stavoren, a photographer who operated a studio out of the top two floors of a downtown building. On the roof Van Stavoren kept a solar enlarging camera, a recent invention housed in a tapering rectangular metal box as large as an automobile. This contraption was designed to counterbalance the era's less sensitive silver-emulsion photographic paper. Via a huge condensing lens, it focused a concentrated beam of sunlight through a negative and onto photographic paper—or projected it onto canvas for tracing before painting. An attendant kept constant watch, frequently adjusting the two steering wheels on the side to keep the telescope aligned with the restless sun. One wheel raised the camera and the other turned it westward. Van Stavoren needed an assistant who was smart enough to perform the job but young enough to tolerate the boredom of such a tedious and repetitive task. If the boy's attention wandered, the camera's focused point of hot light moved to the paper's wooden frame and began to burn it—and risked burning down the building. Van Stavoren's last few assistants had fallen asleep at the job and started fires.

Young Edward did not. "Through summer's heat and winter's cold," wrote Barnard later, he "stood upon the roof of that house and kept the great instrument directed to the sun." Van Stavoren's camera was one of the largest in existence and was named Jupiter. In adulthood Barnard liked to point out that his first astronomy-related job involved attending to a mechanism named after the planet that would most preoccupy his later work. "Since this great instrument's main duty was to do obeisance to the sun, what name more appropriate than that of one of the planets?"

To keep himself awake and avoid boredom, Barnard studied the star to which he and his machine paid homage on a spinning planet millions of miles away. Knowing that the sun reached its highest altitude at noon, Barnard determined that this time of day had arrived when he no longer needed to turn the vertical wheel to raise the camera any higher. He also marked the shadow of a chimney as a guide. "But I was surprised soon to find," he wrote decades later, "that neither of these signs agreed for any length of time with the noon ringing of the bell in a Catholic church near by." Sometimes his own determination of noon, based upon the sun's height and the chimney shadow's length, arrived before the noon bells and sometimes after, "the difference sometimes amounting to a considerable fraction of an hour."

Years later Barnard learned the explanation for this discrepancy between the sun's actual position and the official hours of the day. The reason is akin to why, despite the goofy claims of astrologers, zodiacal signs seldom coincide with the sun's visit to the actual constellation. Because Earth's orbit around the sun is elliptical rather than circular, the two are not always the same distance apart. They are closer during winter and farther apart during summer. (This difference reminds us that the greater warmth of summer results from the axial tilt of the planet rather than from its proximity to the furnace.) For this reason, from our position on Earth the Sun seems to move across the sky more quickly during the winter and more slowly in summertime. Ordinary clocks can't accommodate the variability. As a result human beings have invented the concept of *mean time,* an average of the Earth's circuit of the sun broken down into equal increments despite the seasonal variations of uncooperative reality. Astronomers describe the difference between a clock's high noon and the sun's actual zenith with the evocative phrase *the equation of time.*

I wield the flail of the lashing hail,
    And whiten the green plains under,
And then again I dissolve it in rain,
    And laugh as I pass in thunder.

—Percy Bysshe Shelley, "The Cloud"

Written on the Sky

Since sunrise we have been looking at the sky and the sun as if nothing ever came between them and us, but often our view of both is blocked. Other than the moon (and we will look at eclipses tonight), the largest object ever to cast its shadow on us is a cloud. Anyone who has flown above a cloud bank knows that even up close these ephemeral vapors can look as solid as a mountain range—peaks, valleys, highlights, shadows. It wouldn't be surprising to see Zeus standing with raised thunderbolt in hand.

Flying into a cloud is a surreal experience. Hardheaded realistic technology encounters a barely tangible mist and the world below disappears—just as the plane vanishes when viewed from the ground. Think of all the tens of thousands of years of earthbound feet supporting a brain that looked upward and projected stories onto the sky. Clouds are the Valkyries' steeds and Apollo's flocks; the habitat of Renaissance putti, Hebrew seraphim, Chinese dragons, and Christian angels. No, they're actually just clots of tiny water droplets.

To varying degrees around the world, from tropics to poles, the sky's blue dome is populated with an ever-changing panoply of clouds. They swaddle our globe in a protective layer, screening us from sunlight and filtering it. Not the random agglomeration that they appear, clouds form in obedience to the same strict laws that guide other natural processes. We've looked at the physics behind the rich cloud colors at sunrise, but what about clouds during the rest of the day? They bring rain and hail and thunderstorms, hurricanes and tornadoes. They change literally from one moment to the next over our heads, shaping and foretelling our experience for the next few hours or days. The gray winter overcast in western Pennsylvania, the tumbling capricious skies that remind Londoners to always carry an umbrella, the vagrant clouds that accent the endless blue above the medieval walls of Dubrovnik—these variations remind us that clouds serve as our bellwether to the atmosphere's moods, from benign to dangerous. Millennia ago farmers, hunters, and especially fisherfolk and sailors learned to read cloud signs early in the day and adjust their plans accordingly. For centuries, "Red sky at morning / Sailors take warning" was more than just a folk rhyme.

Many urbanites in the twenty-first century, in contrast, notice a day's varied cloud displays only when trapped in a stadium or lounging on a beach. The rest of the time we get our update on the next few hours' likely weather from television or radio, while dressing for work or while commuting. Distantly, without thinking about it, we have witnessed the gradual evolution of the meteorologist's symbols on television, the technological climb from magnetic cloud icons clinging to a metal outline map to electronic tornadoes that whirl over a tilting three-dimensional satellite image. In our minds these symbols may have become more present and more potent than the reality behind them. Advancing cloud formations—and the changing air they herald—are monitored by Doppler radar, their likely behavior forecast days ahead and constantly revised. The housebound stay riveted to storm porn on the Weather Channel and forget the usually benign panorama of clouds above their own ceiling.

Admonished to keep it simple, meteorologists on television talk about the weather without employing many technical terms, but most people know a few of the words used to describe clouds, if only the three basic shapes: cumulus, cirrus, stratus. These identifying names focus our attention on the specific phenomenon at hand, on the atmospheric conditions that made it possible and the likely next development, making each more fully present to a word-conscious ape such as ourselves. Naming is always the first step in science. The word *cloud* has the rounded cartoon simplicity of *bird* or *tree*—a child's-drawing vagueness—but for those familiar with a few basic shapes, a term such as *cumulus* possesses the image-generating specificity of *eagle* or *willow*. Aboriginal cultures' many names for phenomena that we lump into a single designation (rain, snow, wind) reveal how much more precisely they observed these events than most of us are now capable of doing.

But clouds have not always worn the names mentioned above. Only two centuries ago the very idea of identifying and classifying clouds, of assigning names to varieties and the hybrid shapes between them, was revolutionary and outrageous. To look into this important and little-known development in the history of science, we must go back to the second year of the nineteenth century, to a cold December evening in London, where a shy thirty-year-old Quaker chemist named Luke Howard conquered his disabling stage fright to deliver his first lecture on a topic that had preoccupied him since childhood—the beautiful forms of clouds and the value of a system for observing them.

＊

The British nonfiction writer Richard Hamblyn, in his superb book *The Invention of Clouds: How an Amateur Meteorologist*

Forged the Language of the Skies, wrote of Howard, "There are certain individuals, it seems, who are called by the vastness of the elements, and who return an answering look, upwards or outwards, into the distant views that spread themselves out before their gaze." The social response to such gazing and answering depends upon the time in which it occurs. Howard's public talks took place during the great era of natural history as popular hobby. More than a century before movies and radio—much less television—and decades before even the linotype and other inventions that made possible a reliable daily newspaper, the public lecture was considered entertainment as much as education.

In the late eighteenth century, Enlightenment thinkers from Denis Diderot to Thomas Jefferson emphasized the importance of reason and science in studying both the natural world and human culture. One of their legacies—along with the Industrial Revolution and the long-overdue secularization of education—was visible in the popularity of natural history as a hobby. Proper young ladies in voluminous skirts assembled fern-hunting parties after church; bewigged old men argued over marine fossils exhumed on mountains hundreds of miles from the sea. Crowds filed into chilly lamplit halls to learn about recent discoveries regarding the night sky and the nature of oxygen, the invention of the galvanic battery and the steam engine, and the mysterious revelation that lightning must be a form of electricity. Charles Darwin was a product of this era.

Consequently a boisterous crowd attended Luke Howard's lecture in 1802. Dressed in a plain black Quaker coat and a high white collar, Howard shuffled the handwritten pages of his lecture, "On the Modifications of Clouds," and began to read aloud: "Since the increased attention which has been given to meteorology, the studies of the various appearances of water suspended in the atmosphere is become an interesting and even necessary branch of that pursuit." He went on to explain that, driven partially by practical concerns and partially by a passionate aesthetic appreciation for the phenomena involved, he had formulated a system of identifying clouds by sorting them into categories. He was applying something akin to Linnaeus's binomial nomenclature to the inanimate world. First Howard taxonomically sorted clouds into three umbrella groups, the names of which were both practical and vivid. Moving upward through the atmosphere, away from the surface of the Earth, he identified cumulus, stratus, and cirrus. His proposed names were all coined from Latin. They were admirably precise and simple and thus capable of extensive modification. *Cumulus,* from the Latin for "heaped," was his term for the familiar white fluffy clouds of summer, with rounded tops and flattened bases. Blanketlike layers of clouds he called *stratus,* "layer." It may be helpful to think of these first two in terms of other words derived from the same Latin roots: cumulus accumulate and stratus are stratified. Howard used a word meaning "hairlike" to describe his third major category, *cirrus.* He borrowed another word, *nimbus,* "rainy," and used it as a combining form with other terms, giving us the now familiar *cumulonimbus* for a thundercloud. He described the many transitional and other forms with new hybrid words such as *cirrocumulus* and *nimbostratus.*

Luke Howard's system worked so well that it was almost immediately acclaimed as a brilliant theoretical perception and an equally brilliant practical method for recording and understanding atmospheric phenomena. Soon charts were being printed in many countries, providing forms on which daily observations could be noted. Rival systems modified or even challenged Howard's system, of course, but in time they fell to the wayside because his was so lucid and so helpful. In our own era, atmospheric scientists have established an international system that identifies ten cloud forms, and in their names you will see the Howard legacy surviving and flourishing two centuries after its invention. Scientists divide clouds into three levels in the atmosphere. Each group is determined by the height of the base of the cloud, where the cloud ceiling begins.

The lower level includes those clouds whose base is no more than two kilometers (1.2 miles) above the ground. At this altitude, which in atmospheric terms is just over your head, we find the white, cartoon-friendly cumulus of sunny summer days, as well as layered stratus rain clouds, and the merging of the two called *stratocumulus.* The middle level of cloud formations includes those whose base is between two and six kilometers (between about 1.2 and 3.7 miles) above the ground. Here you find—with the first two prefixed with a Latin word for "height"—*altocumulus, altostratus,* and *nimbostratus.* In the empyrean heights, towering above their fellows, are cirrus, the wispy strands of clouds formed of ice crystals, as well as two more variants: *cirrocumulus* and *cirro-stratus.* In case you're wondering where storm clouds disappeared to in this list, they form a tenth category all to themselves. When you look up cautiously at a towering thunderhead and it seems to rise above the world like an ocean liner beside a dinghy, you are accurately appraising its colossal size. *Cumulonimbus* deserve a separate listing because they rise grandly through all three levels of the atmosphere.

The story of Luke Howard is about the magic of naming. When he looked at the sky and humbly began to sort out its patterns of behavior, when he named the kinds of phenomena he saw above him, he changed the way we see the world. Suddenly the capricious heavens appeared knowable. "Weather writes, erases, and rewrites itself upon the sky with the endless fluidity of language," wrote Richard Hamblyn; "and it is with language that we have sought throughout history to apprehend it."

The great age of such independent, amateur science is over. Science moves so quickly now and has become such a powerful force for change because it is international, well-organized, endlessly cross-referenced, and ultimately (despite competitiveness or the occasional constricting ideology) self-correcting. But it will never outgrow the legacy of such men as Luke Howard, who merged their curiosity, industriousness, and delight in nature into a tool for strengthening our understanding of how the world around us operates on a daily basis. Airplane flights, ships at sea, hurricane monitors, children planning weekend softball games—all owe this Quaker cloud-lover a lasting debt.

Artificial Clouds

As if the diversity of natural clouds was not enough, human beings have added their own variety. In many countries, and almost anywhere in the United States, it is difficult to look at the sky for long without observing one of the twentieth century's many dubious innovations: artificial clouds. Everywhere you turn you can see the white line of a jet *contrail,* a recent word shortened from *condensation trail.* These artificial lines crisscross the blue sky, with consequences that go far beyond annoying landscape photographers and sneaking into the background of historical movies. (Stanley Kubrick's 1960 Roman epic *Spartacus* provokes as many laughs for its anachronistic jet contrails floating above battle scenes as for Kirk Douglas's very 1950s haircut.) Often contrails are beautifully lit with sunset colors or gleam overhead as a pure white scratch long after the sun has gone down, but most of their effect on the world below them is negative.

Contrails provide an observable miniature version of how clouds in general work. Their brief history, which coincides with meteorology's coming of age and is therefore well recorded, helps scientists monitor the role of clouds in atmospheric warming and cooling. Invisible water vapor hangs in the air around us, constituting anywhere from 36 to 70 percent of the greenhouse effect that helps prevent heat from radiating away from the surface of the Earth and into space. Gathered into cloud form, water droplets absorb heat from the surface and reflect it back down, resulting in the warming effect of a cloudy night as opposed to a clear one. But clouds are also simultaneously preventing direct sunlight from hitting the surface by reflecting it back into space—creating, from a distance, the splendid bright blues and whites of the Earth as seen from space. Most low-level clouds, such as banks of cumulus on summer afternoons, cool the surface more than they heat it.

But water in its solid state behaves differently from its gaseous form. The thinner versions of cirrus clouds, formed of ice crystals, let a great deal of sunlight pass through to the ground, but they prevent it from radiating back into space. Thus their overall effect on the atmosphere is to warm it. Contrails not only form artificial cirrus clouds; their own ice crystals disperse at high altitudes and serve as nuclei for the further formation of clouds as they gather water molecules. The contrail is basically an induced cirrus cloud, formed because jets fly at impossibly cold heights where the temperature is no more than minus 70 degrees Fahrenheit. Jet engines expel exhaust particles, including moisture. At this height, air contains little humidity and will quickly absorb the new moisture from the exhaust. It lingers only when the air is more humid. Contrails form thin layers of cloudlike particles in the atmosphere, slightly reducing Earth's surface temperature by shading us from incoming radiation during the day and slightly raising it by preventing nighttime radiation of heat back into the atmosphere. Generally, as with natural clouds, the daytime cooling effect outweighs its nighttime opposite.

This evening we will see contrails adding to the show at sunset, but there were none visible at dawn. Jets fly mostly during daytime, and the majority of the contrails of the fewer nighttime flights scatter before sunrise. Like clouds that form naturally, contrails don't maintain one shape for very long; they dissipate at various speeds, depending upon wind velocity and other factors. For this reason, the timing of jet flights during the day can affect overall warming. Atmospheric scientists observed that in September 2001, during three days in which airlines were grounded after the al Qaeda terrorist attacks in the United States, daytime temperatures rose measurably and there was a corresponding drop in nighttime temperatures. Minus jet contrails, the average temperature in the United States during these three days was 1.8 degrees higher than normal. In the years since, many scientific studies have demonstrated more conclusively what this first paper merely hinted at: that contrails directly contribute to surface warming. One study reported that, although average humidity at the altitude of cirrus cloud formation was shown to be constant between 1974 and 1994, detailed satellite photographs and other weather data confirmed that during this period there was a dramatic increase in cirriform clouds. The fluctuations seem to correspond directly to the volume of airline flights. There is also a great deal of local evidence from around the world about the atmospheric effects of jet flights. Vandenberg Air Force Base on the California coast, for example, sometimes experiences an artificial twilight because of the vast amount of exhaust particles propelled into the air during test launches of ballistic missiles or other high-flying aircraft.

In 2006 British scientists reported that the one-quarter of air flights that occur between 6 p.m. and 6 a.m. produce between 60 and 80 percent of contrail-induced warming. This discovery suggests that, with global warming an ever more urgent concern, responsible travelers might prefer daytime flights when the option is available. The British scientists also discovered a seasonal variation. Although fewer jets take to the air over England during cold months, increased humidity during this time doubles the production of contrails. It looks as if the best time for an environmentally responsible flight would be a daytime winter one. Researchers have also recommended that planes begin flying lower, below the cold heights in which these artificial clouds form and begin to shadow the ground below.

Gunfight in Abilene

As the clouds disperse and float over the horizon, we realize that the sun's harsh spotlight is almost directly overhead—blurred only by the gauzy interweaving of jet contrails. It's time for lunch. As you leave the office you realize that during your hours indoors the air has become even warmer than you expected. The sunlight is so bright you squint and regret leaving your sunglasses behind. Now that we have reached the middle of our day, let's take a cautious peek at the noonday sun and at how our relationship to this time of day has altered over the last century or two.

One of the standard images from old westerns is the gunfight at high noon. In this persistent myth of the Wild West, the hero and his black-clad adversary face each other down the long dusty main street of Abilene or Deadwood or Tombstone. Ordinary folk cower on the sidelines, watching from behind prop-shop swinging doors and candy-glass windows that will soon be shattered by a stray bullet. The duel is held at "high" noon so that the sun's glare will be in neither man's squinting eyes: part of the myth of fairness, that the outcome of the fight ought to depend solely upon speed of draw and accuracy of aim. In the movies, sweat beads on furrowed brows. The sun beats down. The shadows around the gunfighters are short—unless, as occasionally happened, the continuity editor fumbled and inserted a close-up or a reshoot without matching the lighting angle.

The 1952 film that helped preserve the phrase, Fred Zinnemann's bleak and cartoonishly stylized *High Noon*, actually employs none of these clichés. Rather than an image of dwindling shadows as the sun climbs higher, viewers take away the memory of inexorable clock hands. Halfway through the twentieth century, the sun's high noon of earlier westerns has been replaced entirely by the clock's high noon. Laconic, tight-lipped Gary Cooper and his enemies don't face each other in a quick-draw duel on Main Street; their gunfights take place in the livery stable, around building corners, through broken windows. A fair fight was never an option. In this film the minute and hour hand converge to point straight upward to the moment when the gang leader returns from prison with vengeance on his mind. For the rest of us, in office or home, the clock hands meet and point straight up toward a sky whose actual noon, somewhere above the ceiling, is far less important to us than the clock's official version.

It seems appropriate that this vivid image emerges from a film portraying the end of the Wild West and the importance of clocks in train schedules. Beginning in the 1840s, as train tracks stretched across the North American continent, telegraph lines paralleled them. Trains could not have become so efficient without the telegraph that permitted station agents to forecast arrival times, schedule track use, and report delays. Telegraph was the fastest form of communication yet invented, the first in all our history that did not rely upon some form of locomotion, be it four-legged, two-legged, or steam- or wind-powered. Not surprisingly, natives and brigands cut the lines to level the playing field. Telegraph operators assumed oracular importance; their offices became public gathering places for people urgently seeking updates—about weather, natural disasters, battles, elections.

"And around noon," in the words of historian John R. Stilgoe,

> they drifted to telegraph offices for the time signal, the announcement of precise noon that flashed across the United States once each day, the single click of the telegraph sounder that enabled bankers and shoemakers to synchronize their watches with those of railroad conductors and engineers, to see how accurately their timepieces ticked, the single click that killed rural time, small-town time, personal time.

✳

In our era, of course, trains, planes, workdays, weather reports, and rush-hour radio programs all require to-the-minute schedules. Most adults in the industrialized nations wear a watch, the majority of them digital. The older-style analog watches, with their steadily moving hour and minute hands—and the alarmingly fleet second hand, frantic index of our mortal lives—provide a sense of flowing time. Digital watches, in contrast, pin down a single electronically calculated instant but provide no visual reminder of elapsed time or the amount of time between this moment and lunch, the end of the workday, bedtime, whatever. Preschool and elementary school teachers report that it is much easier to inculcate an awareness of passing time with an analog clock; quarter till three and half past two make no visual sense on a digital clock.

Once the smallest imaginable unit of time, the second is now subdivided into smaller pieces—nanosecond, millisecond, New York second—just as the atom has been unseated from its diminutive throne by upstart quarks and other impossibly tiny particles. Mobile telephones and travel alarm clocks, automobile phones, computers: each automatically synchronizes itself with an internationally agreed-upon arbiter of time. Many people still set their clocks ahead a few minutes to pretend to themselves that they have more time left before work than the clock admits. Meanwhile the Earth turns on its axis, and to our earthbound eyes the sun appears to have begun its slow descent toward the western horizon.

Firmament

The sky constitutes half of our visual field, but mostly we keep our eyes lowered—and half of our universe goes unnoticed.

—Chet Raymo, *Natural Prayers*

After

Noon is firmly established as the pivot of the day. We have climbed the mountain and are now skiing down the other side. Dividing the day at noon led to our current designations of *a.m.* and *p.m.*—*ante meridiem* and *post meridiem,* which translate literally as "before noon" and "after noon." Like *nonfiction* or *unkempt,* the word *afternoon* defines itself in terms of what it is not, but the earlier half of the day has its own designation. The now quaint word *morn* has been around since the thirteenth century, and the extension *morning* from a few centuries later. Originally it was a parallel construction along the lines of *evening,* itself an adaptation of the older terms *even* and *eve* for the period following darkness but prior to midnight before a certain day. This usage remains, of course, in Halloween—originally Hallowe'en, Hallowed Evening, All Hallows' Eve, the period after dusk and before the midnight that officially begins 1 November, All Saints' Day.

Shredding and slicing, dividing and subdividing, the clocks of Harley Street nibbled at the June day, counselled submission, upheld authority, and pointed out in chorus the supreme advantages of a sense of proportion, until the mound of time was so far diminished that a commercial clock, suspended above a shop in Oxford Street, announced, genially and fraternally, as if it were a pleasure to Messrs. Rigby and Lowndes to give the information gratis, that it was half-past one.

—Virginia Woolf, *Mrs. Dalloway*

The Flapping of a Black Wing

How did it get to be afternoon already? Has there ever been a human being who lived a few decades without muttering about how quickly time flies? No doubt elderly (that is, thirty-year-old) Neandertals noticed that *tempus fugit* long before Romans condensed daily experience into an aphorism. Our similar habit-bound days blur together until we can't remember whether we took our thyroid medication this morning or yesterday or last week; we carry seven-day boxes to remind us. A box of birth control pills outlines an entire month. The primary value of secular holidays is that they force one day to stand out among the whirl. Philosophers, comedians, and tipsy birthday celebrants have all proposed theories about why time seems to move increasingly swiftly as we grow older. But the most disconcerting rationale is not a theory. It is the undeniable realization that every day we live constitutes a smaller percentage of the accrued experience with which we awaken each morning, and therefore seems proportionately a smidgen quicker and smaller than the day before.

It is bizarre to think about how many times the sun has passed over our heads since we were born, how many times we have opened our eyes to a new day—especially if we dwell upon how many of those days we have spent in ways that we later regretted. In childhood the sun seems to crawl through the sky during schooldays, leisurely climbing to zenith and even more lazily sliding down toward the afternoon departure bell. Rush hour seldom rushes by. As you wait for news about a family emergency, the moon seems stationary outside the window, and then when you look away it hastily drops to the horizon like a stage effect. Yet in retrospect our days blur together. It is worth thinking about this telescoping of days over time, in order to develop a mental image of how days add up to weeks and months and years and decades and lifetimes—and even those periods longer than lifetimes, centuries and millennia. How else are we to acquire a cautionary idea of the innumerable spins of our planetary top over the endless time before we were born and after we die?

It is a dizzying image, and we may be fortunate that we can witness such a distillation only in imagination. A good place to do so is the first book by H. G. Wells, his little 1895 fantasy of class war run amok, *The Time Machine*. It's a slap-dash affair, with young Wells unable to decide if he wants to become Jonathan Swift or Jules Verne when he grows up. The author himself later dismissed the novella, in which action regularly screeches to a halt for Fabian Society lectures, as an "undergraduate" attempt. Yet its vivid imagery has had a lasting effect on science fiction. Most powerful of all is Wells's method of dramatizing the passage of time in a scene that foretold cinema montage. Because it so nicely embodies our themes today—and because we are all time travelers, and the day is our vehicle—it deserves spacious quotation, from the Traveller himself, who has climbed aboard his time machine:

> I pressed the lever over to its extreme position. The night came like the turning out of a lamp, and in another moment came to-morrow. . . . To-morrow night came black, then day again, night again, day again, faster and faster still. . . . As I put on pace, night followed day like the flapping of a black wing. . . . The twinkling succession of darkness and light was excessively painful to the eye. Then, in the intermittent darknesses, I saw the moon spinning swiftly through her quarters from new to full, and had a faint glimpse of the circling stars. Presently, as I went on, still gaining velocity, the palpitation of night and day merged into one continuous greyness; the sky took on a wonderful deepness of blue, a splendid luminous color like that of early twilight; the jerking sun became a streak of fire, a brilliant arch, in space; the

moon a fainter fluctuating band; and I could see nothing of
the stars, save now and then a brighter circle flickering in
the blue. . . .

In the 1960 George Pal film of the book, actor Rod Taylor
climbs onto the rococo vehicle that looks like a sleigh for an
eccentric Santa Claus, sits in the upholstered barber chair
that serves as pilot's seat, and pushes forward the lever that
activates the crystal power source. Behind the seat spins a
large solar disk bearing 365 knobs to mark the days in a year.
Colored lights flash on a control panel, above a screen efficiently adding days like an odometer (and slyly bearing the
legend "Manufactured by H. George Wells"). The special
effects crew conveyed the passage of time through various
sequences—a tree leafing out and growing apples, a melting
candle beside a clock with spinning hands, a laughably
mechanical snail racing across the floor. By far the most successful image was the simplest, which was drawn directly
from Wells's novel and returned to the daily experience of
each of us: seen through the laboratory's skylight, clouds and
sun and moon race across a sky that flashes light and dark
like a strobe.

A Strong Metal Bowl

Writing about torture in his classic essay "The Hour of
Poetry," the great art critic John Berger—always returning
like Rilke to a moral aesthetic of witness and testimony—
includes a poem that complains that, despite the parade of

atrocities under it, "the sky never changes." Actually, despite nature's maddening "No comment," despite the moon's laissez-faire illumination of both courtship and battlefields, the sky changes constantly. It is a theater of endless drama, even when we haven't revved it up to time-travel speeds that make the sun blink on and off. But it is true that it is always comfortingly present and in time returns to its default mode of blue serenity.

Look up. Raise your eyes. Gaze heavenward. Sunlight, moonlight, rain, snow—these gifts or punishments are bestowed upon us from above. Most cultures associate the sky with deities. Verticality is certainly prevalent in our imagery, perhaps as much from our own bipedality as from the presence of the sun. We climb the corporate ladder and achieve summits in our careers and reach for the stars, look up to heroes and look down on the less advantaged. The architecture of cathedrals pulls our gaze upward.

Whether depending upon the sun or moon to illuminate their path, preindustrial peoples always looked to the sky. Superior beings reside there. Where else could be worthy of them? The omnipotent sun lives in the sky, reigning until the arrival of its nighttime regent. Light itself comes from above. Heaven can be generous with vital rain or withhold it for reasons known only to itself, but early religions assumed that bad weather, like other afflictions, indicated a parental disapproval over some act we had performed or omitted. Storms develop in the sky and pummel trees until they flail in submission. Vengefully the heavens fling lightning to earth, demolishing trees and starting fires and occasionally stealing the life of a human being. Even before we faced acid rain and disintegrating artificial satellites, we knew that the sky may rain treasures or plagues. Sky Woman, in an Abenaki creation myth, falls to earth through a hole in the sky caused by the uprooting of a celestial tree. Indeed, our imaginative response to gravity's pull has turned falling itself into a metaphor. We say that civilizations rise and fall and we imagine a golden age before the Fall of Man. Lucifer is thrown out of the celestial realm and plummets to the mundane as surely as Elijah rises heavenward.

Superior beings reside in the sky because it is beyond us. Surely not even the tallest peak could lift you high enough to crowd the gods—although Yahweh does thwart the ambitious architects of Babel, who yearn to construct a tower "whose top may reach unto heaven." Mountains have always been considered sacred because they take us far enough into the sky to permit us to see the setting of our everyday lives in both literal and figurative perspective. Greek gods scheme and bicker on Mount Olympus; Kiowa deities are raised into the sky by an igneous outcropping growing higher and higher until it forms the Devil's Tower in Wyoming. To catch the ear of the gods, we get as close as we can to what Czeslaw Milosz called "the inhabited / Classical sky." The Pyramid of the Sun and the smaller Pyramid of the Moon, amid the ruins of Teotihuacán, are not angular pyramids as much as pyramidal imitations of the surrounding mountains. Ancient ziggurats aspired heavenward as surely as mosques do now.

※

Naturally human beings have always asked themselves the question about this crucial domain, "What *is* the sky?" Without an understanding of atmosphere or spectrum, many early skygazers regarded the heavens as solid and unchangeable— that is, at least the reliable blue background of the traveling clouds and sun. The sky seemed a perfect sphere encasing below it the mucky and decay-prone earth. There are versions of this idea in many sources, including Homer, but let's explore a particular example from the Bible.

The English word *firmament* boxes up the tradition of the sky as a sphere so tidily that it invites an etymological unpacking. Like most such stories, this one digs ever more deeply into the history underlying everyday life, so we will move backward in time. The current word *firmament,* meaning the vault of the heavens, dates from thirteenth-century Middle English, but it became widely known and biblically significant only after the publication of the King James Bible in 1611. The king's translators adapted the word from the Ecclesiastical Latin *firmamentum* employed in the fourth-century Vulgate Bible. The Vulgate (from the Latin *editio vulgata,* "common version") was created at the behest of Pope Damasus and under the aegis of the biblical scholar Jerome. It was the first comprehensive and authoritative Latin version of other translations then in use, especially the Greek Septuagint (named, in turn, for its seventy-odd translators).

Jerome's team chose *firmamentum* as the translation of the Greek word *stereoma;* both meant "solid and strong." Earlier the scholars of the Septuagint had chosen *stereoma* to translate the Hebrew word *raqia',* which referred to a thin layer of beaten metal. In Hebrew mythology the dome of the heavens was perceived as a literal dome, a hemisphere arcing over the flat earth and supported at its perimeter by mountain pillars. In the Bible Job says specifically, "He hath compassed the waters with bounds, until the day and night come to an end. The pillars of heaven tremble, and are astonished at his reproof." Later, during his party-line rebuttal to Job's complaints about how the cosmos is run, Elihu asks, "Hast thou with him spread out the sky, which is strong, and as a molten looking-glass?"

It was this steadfast dome that compassed the waters of heaven and separated them from terrestrial waters. Hence God's fiat on his second day of work:

> Let there be a firmament in the midst of the waters, and let it divide the waters from the waters. And God made the firmament, and divided the waters which were under the firmament from the waters which were above the firmament: and it was so. And God called the firmament Heaven.

It was thought that the Deluge resulted from God opening windows in the firmament that permitted the waters of heaven to sluice down and rejoin the waters of earth. God might open these windows as he might express displeasure in any other way, but when the old boy was in a better mood the sky was considered strong and trustworthy and solid.

But what if this reliable sky is *not* held fast above our heads? The question seems a reasonable next step, and many people have asked it. If the sky is solid and somehow held up, then like everything else in our experience it might not stay up. "The sky is falling!" cries the alarmist hen in the story of Chicken Little. In Aztec mythology, the Fourth Sun of creation—the age preceding the present time—ended when the land was flooded while simultaneously the sky fell down. The Fifth Sun, the time in which the Aztecs recorded these stories, began when Ometeotl's sons raised the sky back to where it belonged.

A different kind of movable sky shows up elsewhere. The Kadaru and Nyimang peoples of the Nuba Mountains in Africa tell related stories about a time when the sky was much closer than it is now. The Kadaru myth, in a charming image, tells of their ancestors reaching up and cutting off pieces of cloud to eat like cotton candy. For some reason, an angry woman stabs the sky with a long-handled stirrer. The sky not only withdraws to a safe distance but thereafter withholds rain except during a brief annual season. In the Nyimang version the woman stabs the sky with her stirrer, but she does so because it crowds so low that the people must go around bent over. They even have to hold their hands so low over cooking pots that their fingers burn.

As I walked down toward Newcastle, the sun hammered me. I held my face up, said,

  "Roast me, yah bastard."

—Ken Bruen, *The Guards*

A White-Hot Arrow

Although much of weather is caused by the interaction of the earth and the sun—the tilted planet turning, the land and water heating unevenly—weather is much too large a topic for us to explore at length in this book. But let's glance at one aspect of it that we experience every day, here under the metal dome of the sky. Just as autumn is warmer than spring, so is afternoon warmer than morning. During the winter and during the night the earth cools off; heat, vernal or matutinal, requires time to build up. (Everywhere we turn we find the history of sun-worshipping. The very adjective *matutinal,* referring to the morning, echoes a Roman goddess of the dawn, Matut or Matuta.) Naturally the temperature—unless a warm front is bullying its way into the neighborhood— can't reach its zenith until after the sun does. When left to our leisure, we tend to take our time indoors on cooler days, until the air has absorbed some energy from the feeble sun.

As far as nature is concerned, there's no rush; this time of day doesn't hurry. Sunrise and sunset race ahead, changing every second. Even dawn and twilight—the day's early spring and late fall—always move forward; you have to keep your eyes on the stage or you miss part of the show. The middle of the day, however, like the middle of the night, like midsummer and midwinter, seems to tarry. Without a clock to guide us, the time from midmorning to midafternoon at work can dawdle with nowhere to go. We linger over lunch or stare out office windows and daydream until time ceases to move, or at least until a ringing phone jolts us from the trance into which too many office afternoons disappear.

※

The occasionally unpleasant heat of a summer day is one of many topics upon which Shakespeare and Cole Porter concur. "Sometime," complains the sonneteer, "too hot the eye of heaven shines," and Ella Fitzgerald has to agree: "It's too—darn—hot." The sun, even for those living far from the equatorial furnace, is not always friendly. Shakespeare's countryman and heir in spendthrift imagination, Charles Dickens—child of London fogs and shadowed alleys— opened his magnificent late novel *Little Dorrit* with a horrific vision of the sun's glare in coastal France: "Marseilles lay burning in the sun. . . . The universal stare made the eyes ache. . . . Blinds, shutters, curtains, awnings, were all closed and drawn to keep out the stare. Grant it but a chink or keyhole, and it shot in like a white-hot arrow."

But the sun's anger can get much worse. What if you have no recourse, no shade to flee toward—even no chance to answer your own urge to seek the shade? In the early twentieth century, black inmates on Texas work-farm prisons would labor all day in literally blistering heat. Some of those who didn't die of sunstroke or more human abuse often sang a work song that stated even more baldly than Dickens how fierce the sun can be at times, and how easy it is to imagine the world ending in fire rather than ice:

Go down, old Hannah,
Doncha rise no mo',
If you rise in the mornin',
Bring Judgement Day. . . .
If you rise in the mornin',
Set the world on fire.

※

Most of us never experience such hardship and, except at the height of summer, have no quarrel with the sun. Up to a point, a glow from being outdoors brightens most faces and looks healthy and attractive. Dermatologists constantly remind us that any change in skin tone reports a measure of damage, but we are diurnal mammals and our bodies require sunlight. Vitamin D, which assists the human body in absorbing calcium and phosphorus for bone formation and other processes, already exists in skin, but it is activated only by sunlight. Consequently in our indoor era the United States and other nations fortify cow's milk with vitamin D to help prevent rickets.

Until recent centuries, most people spent their days outdoors. Whether tilling the ground for master or self, they were exposed to the sun's fire through long days of backbreaking labor. In time people of European ancestry began to value pale skin as evidence of a less toilsome life, of days spent loafing, shaded by privilege. As more work moved indoors, only the lower classes remained out there in the sunlight. Hence such dismissive terms as *redneck,* and the less offensive, more recent description *farmer's tan.* (The former may also trace partially to pellagra, a disease caused by niacin deficiency, whose many side effects include a dermatitis that can turn the skin of the neck thick and red, and which was associated with poor whites in the rural United States.) But then came forklift drivers and elevator operators, and then business management hierarchies and personal computers, and it seemed that even the lower classes were working indoors. Their proletarian tans began to fade. By the mid-twentieth century the tan, no longer a badge of blue-collar employment, was promoted to an emblem of freedom from toil. Cary Grant worked on his tan almost every day, spending his holidays from filmmaking on the nearest sunny beach, holding a reflective card under his chin to reach nooks hidden from the light. To the non–movie star, returning to the office with a tan meant that you had spent your holiday (or even your weekend) jet-setting to Aruba rather than scrubbing the toilet. It didn't matter if you had been to Far Rockaway rather than Aruba; the effect was the same. Tans were In; pale was Out.

Then the pendulum swung again. Beginning in the 1970s, scientists began to report that the ozonosphere, the outer-atmosphere layer of ozone that protects the earth's surface from solar ultraviolet radiation (which we will examine shortly), was growing less dense, less protective. Our home was losing some of the armor that protects it from the sun's ferocity. Soon dermatologists and oncologists were insisting that the best protection for your skin—and the best antidote to that dreaded terminal disease, "aging"—was sunblock. As you may recall from the beginning of Phaethon's joyride, even the driver of the sun's chariot advised his offspring to wear protection from its fire. Tanning "salons," which originally were designed to artificially induce tans for people who wanted the cachet without the burden of actual time outdoors, now regularly offer spray-on options: skin dyes marketed as tan-in-a-bottle. We may have learned to be wary of the sun, but we still cherish the twentieth-century emblem of the good life, proof that we are not mere cubicled drones— at least not on the weekend.

What Makes the Wind Blow?

Weather's local manifestations vary, but the regular movements of the Earth and sun, and the air and water currents stirred by them, create variations in temperature and humidity and precipitation throughout the day. Earth's rotation on its axis and its revolution around the sun affect our weather phenomena in two main ways. First, of course, are the warming rays of the sun itself, and their varying impact on different regions as the Earth turns one area toward and then away from solar radiation.

The second crucial influence results from another side effect of this rotation. While Earth spins on its axis, different latitudes experience different rates of rotation. As Eratosthenes discovered in his shadow experiment, the circumference of the earth is about 25,000 miles (40,000 kilometers). He measured longitudinally, around the poles, which is actually a slightly smaller measurement than equatorially because Earth's spin keeps it flatter at the poles, but these differences don't matter for our purposes. Because the planet rotates on its axis once every twenty-four hours, a spot on the equator moves 25,000 miles through space every day—and we're ignoring Earth's revolution around the sun and the sun's own progression through space.

If you live at the equator, even if you spend the day in bed you are moving at a speed of 25,000 divided by 24, or more than 1,000 miles per hour. Obviously latitudes to the north or south of the equator measure smaller circumferences. While an Ecuadorian is zipping through space at more than a thousand miles per hour, an Inuit in Anchorage is moving a little more than half as quickly. Latitude lines are parallel but encompass ever smaller areas as they move north or south. Longitudinal meridians, in contrast, all measure the same length, because they merge at the poles, but the distance between them narrows as you go north or south of the equator.

Variably heated areas at diverse latitudes result in the different speeds at which air masses move over water and land, creating wind and rain and storms. Weather nourishes and sculpts the very ground beneath our feet. It enlivens the skies above with an endless parade of clouds and other phenomena that inspired many aspects of our cultural heritage—and led our imaginations (and eventually our technologies) upward into the sky and beyond.

※

This is the time of day when the wind is likeliest to lose control. The most common time for tornadoes in most parts of the United States is now—late afternoon or early evening. Well over half occur between 2 and 9 p.m. This entertaining quirk of physics is another side effect of the buildup of heat throughout the day during the warmer months of the year. Tornadoes are least likely between sunrise and midmorning, when the surface of the earth and the air above it are at their coolest.

In O. Henry's famous story "The Ransom of Red Chief," the insufferable little boy asks, "Do the trees moving make the wind blow?" Actually, kid, the planet moving makes the wind blow. If Earth's temperature were uniform across its surface, there would be no wind to whip itself into a tornado. Instead, as it turns its face toward and then away from the warming rays of the sun, the amount of solar radiation varies on different regions. This disparity causes air movement. Wind blows from cooler regions to warmer; warm air rises because it is lighter and less dense, and cooler air moves in to replace it. Meanwhile the Earth's spin causes the air to turn too, creating still more atmospheric currents. Consequently, high pressure in the Northern Hemisphere moves in a clockwise pattern and low pressure counterclockwise. As a severe thunderstorm develops, fast-moving wind higher up can meet slower wind near the ground and start the air spinning. Some of these whirling proto-tornadoes sputter out immediately. Under the right (or wrong) conditions, others keep going and grow stronger—and meaner.

A tornado works like a drinking straw. Inside a container, atmospheric pressure is the same on every part of the surface of a drink until you draw air upward through the straw. As you suck on the straw, you reduce the air pressure inside it. The pressure on the rest of the surface remains the same, so the liquid rises upward in the area of reduced pressure. You have created a vortex, like an upside-down bathtub drain, pulling air and liquid upward rather than downward. A close-up view of the bottom of your drink would show everything rushing toward the submerged end of the straw, just the way that a tornado pulls everything into its low-pressure end on the ground.

It is sobering to think that this impressive natural suction results from the Earth's casual daily spin. No wonder meteorologists worldwide quote the same Zen dictum: "Everything is connected."

Policeman Ozone

The weird interconnectedness of nature is nowhere more evident than in the strange story of ozone. Many chemical compounds and forms of particulate matter contribute to air pollution, which is an omnipresent fact of life in the twenty-first century—and has been, in one form or another, ever since protohumans brought home a lightning-ignited branch and tamed fire. A couple of kinds of pollution form and vary in response to the sun's progress across the sky and the consequent daily cycle of warmer and cooler air.

By this time of afternoon, sunlight has been baking the streets and parking lots for many hours, and auto exhaust fumes and other pollutants have been adding up all around us. The fresher, cleaner air of early morning, when you walked out of your home and headed to work, has long since been trashed by accumulating pollutants. The foul breath of our cars and businesses is partnering with the treacherous sun to manufacture trouble.

This time of day in the warmer months, from midafternoon onward, is when ozone pollution reaches its daily peak. The spike on the air monitors' graphs occurs because of the interaction of two chemical compounds whose acronyms sound like Dr. Seuss characters: NO_x and VOCs. NO_x is the chemical abbreviation of nitrogen oxides. The one that most concerns us, nitrogen dioxide (NO_2), is generated almost solely by the combustion of fossil fuels in motor vehicles, in industrial boilers, and in coal-fired power plants. *VOCs* is the bureaucratic abbreviation of volatile organic compounds, a group of troublemakers that includes those gases produced by automobiles and factories, labeled *volatile* because of their promiscuous willingness to form alliances with other compounds. Each day's new NO_x and VOCs become ingredients in the recipe that, when warmed by sunlight, form that ubiquitous element in urban life, smog, which we will address shortly.

This invisible chemical factory in the air is why ozone levels rise during the day and fall at night, and it is why it is healthier to exercise outdoors during early morning instead of late afternoon. One Vanderbilt University study demonstrated that exercise buffs running along the circumference of the university's campus during Nashville's rush-hour traffic, when five lanes of cars idled at traffic lights or crept slowly by, were inhaling as much carbon monoxide and ozone as students who stayed in their dorm and smoked four cigarettes.

In those nations that maintain official standards, environmental agencies collect data about ozone to monitor compliance within limits. When a city or other region exceeds limits, monitors must begin the challenging detective work of isolating the culprits. Although automobile traffic's chemical by-products are the primary contributors to ozone formation, measurements indicate that its levels in cities are sometimes lower than in the surrounding countryside, because valley ecosystems often trap moisture and pollutants and exacerbate ozone levels. In the Smoky Mountains National Park in the United States, ozone levels are higher than in nearby cities such as Knoxville and Charlotte. However, this discrepancy doesn't mean that, as that distinguished scientist Ronald Reagan once claimed, trees cause air pollution. It occurs because the chemical pollutants generated each day mix with hydrocarbons.

The ozone molecule is formed of three oxygen atoms, another of those common cases of a slight variation in an essential element producing its Mr. Hyde dark side. Ozone can be condensed into liquid or solid forms, but both are dangerous explosives. In its pure state ozone is bluish and has a pungent metallic smell. Fortunately we seldom encounter it in its pure state, but we do run into it on a daily basis, without noticing.

Ozone was discovered in 1839. Its name, from the Greek *ozein*, "to smell," was coined by a German-Swiss chemist, Christian Friedrich Schönbein, because of the olfactory evidence of its presence in air after lightning. What he was actually smelling, however, was not ozone itself but molecules released during the sudden chemical changes caused by lightning's intense heat.

In 1865 William S. Gilbert, best known now as the versifying half of Gilbert and Sullivan, actually wrote a poem about ozone. It demonstrates the widespread Victorian notion that the element was scarce in cities and plentiful in mountains. The Scottish peak Ben Nevis (Beinn Nibheis in Gaelic), mentioned in the following verses, is the highest mountain in the United Kingdom.

> But if on Ben Nevis's top you stop,
> You will find of this gas there's a crop—but drop
> To the regions below,
> And experiments show
> Not a trace of this useful ozone is known,
> Not a trace of this useful ozone!
>
> It's because I'm an ignorant chap, mayhap,
> And I dare say I merit a slap or a rap,
> But it's never, you see,
> Where it's wanted to be,
> So I call it Policeman Ozone—it's known
> By my friends as Policeman Ozone!

The amount of naturally occurring ozone in the atmosphere varies, but averages out to about 20 parts per billion. This time of day in the summer, however, the ratio can rise to more than five times as much, and has been recorded at more than 260 ppb. Ozone causes a variety of respiratory ailments when we inhale it into our delicately balanced airways.

It cuts down the body's resistance by damaging the lungs' macrophage cells, which consume bacteria. Furthermore, the bronchioles in the lungs respond to ozone with asthmalike hyperreactivity, creating much the same internal chaos as a viral infection, in severe cases even causing death. Asthma patients suffer most, but many other people experience discomfort and pain and steadily worsening respiratory troubles throughout the warmer months—another way in which the day echoes the year. Ozone also damages plants, undermining their defenses and making them more susceptible to attack from cold weather, insects, and fungi. Levels of only 50 ppb have been proven to reduce the yield of domesticated crops by as much as a sixth.

※

And yes, this is the same ozone that forms an important layer of our atmosphere. There are good reasons why ozone is, as TV meteorologists tend to simplify it, "good up there and bad down here." The ozonosphere, the ozone layer in the upper stratosphere, is a crucial part of our insulation against the random radiation that fries unguarded planets. Most dense at an altitude of thirteen to sixteen miles, it absorbs 99 percent of the ultraviolet radiation that rains upon our atmosphere from the sun—or at least it did so until recent chemical depredations reduced its density and effectiveness.

The ozone layer isn't a waiting army that always guards our planet's border against enemies; instead, like a militia, it forms in response to attack. When the sun's short-wavelength ultraviolet light reaches our upper atmosphere, it divides some molecules of oxygen gas (O_2) into their two constituent atoms. These freed atoms then couple with remaining oxygen molecules to form ozone (O_3), which absorbs most ultraviolet light and keeps it from damaging Earth's surface. The role of oxygen in this process reminds us of a historical tidbit that paleontologists like to point out: before free oxygen formed on the primordial globe, there was no ozonosphere. Life could exist only underwater, where it was screened from ultraviolet radiation by water's density. Scientists estimate that a version of the ozonosphere has been protecting Earth for the last billion years or so—one of those offhand guesstimates that the rest of us automatically translate into "a really, *really* long time."

There are three kinds of ultraviolet light, and the ozonosphere contributes to protecting us from each. It absorbs only a small percentage of UVA. It is UVB, however, that is most dangerous to living things—interrupting photosynthesis, the foundation of food chains in plants; killing phytoplankton, the foundation of food chains in the ocean; and suppressing immune system response and causing cataracts and skin cancer in human beings. And ozone absorbs almost all of the UVB hitting the atmosphere. The ultraviolet light with the shortest wavelength—which makes it lethal enough to use in sterilization—is UVC, and, in partnership with oxygen, ozone absorbs this too.

For decades, volatile chlorofluorocarbons, or CFCs, were blithely produced in the manufacture of refrigerants, aerosol propellants, and solvents, and as by-products of aircraft and nuclear weapons. They are a threat to the ozone layer because they contain, as the first syllable indicates, chlorine. (Other compounds also damage ozone, including chlorine by itself and bromine, familiar to us in the once common insecticide methyl bromide.) Early in the 1930s, the American chemist Thomas Midgley invented Freon, the first CFC, and innocently contributed to endangering his planet's ecosystems. It was also the enterprising Midgley who determined that lead additives would increase performance in combustion engines, a discovery that quickly led to far more dangerous pollution from automobiles. These combined innovations make Midgley, in the words of the environmental historian J. R. McNeill, the man who "had more impact on the atmosphere than any other single organism in earth history." It is difficult to resist mentioning that Midgley accidentally killed himself with another invention, a system of pulleys and ropes that he designed to lift him out of bed after he contracted polio.

Gaseous compounds that include chlorine eventually diffuse throughout the atmosphere and deliver their chlorine atoms to the ozone layer. This is a dangerous situation for us naked apes, as well as for our multitudinous kin, because every chlorine atom exposed to the sun's ultraviolet radiation metamorphoses into a superhero atom that has the power to destroy about a million atoms of ozone. In late 2006, several monitors at the South Pole, including those of the National Oceanic and Atmospheric Administration, reported that the now carefully watched hole over the Antarctic was exhibiting some of the worst depletion of the ozonosphere ever observed. The report stated that ozone in the 14- to 21-kilometer layer (roughly 8.5 to 13 miles up) had been almost entirely destroyed. As so many news programs are finally noticing and reporting, the likely global consequences of this invisible war between natural processes and technological recklessness are terrifying.

The depletion of the ozonosphere, in response to the buildup of greenhouse gases, is also why skin cancer is on the rise, and why sunblock has become our feeble terrestrial backup defense, our private armor to make up for the slackening vigilance of the chemical guardian above us. William S. Gilbert's nickname for this element was more apt than he realized.

A Presumptuous Smoke

Ozone may be invisible by itself, but it contributes to the haze that is part of too many urban afternoons. The primordial (but still evolving) history of ozone reminds us that air pollution is not a new problem. It has existed at least in short-lived, localized, natural forms since before there were human beings to cough in response. Yet, always tampering with the world around us, we are masterful when it comes to throwing nature out of balance. Steadily increasing population has required the combustion of steadily increasing amounts of fossil fuels. For hundreds of years now, the air we breathe has been growing increasingly dirty and dangerous.

The first recorded environmental legislation, enacted by King Edward I in thirteenth-century England, prohibited London merchants from burning coal while Parliament was in session. Like many such prohibitions, it didn't go far enough. In 1661, the English writer John Evelyn wrote a pamphlet about the ongoing problem; its memorable title was *Fumifugium, or, the inconvenience of the air, and smoke of London dissipated.* Evelyn described a day on which "a presumptuous smoke issuing from near Northumberland house, and not far from Scotland Yard, did so invade the Court, that all the rooms, galleries and places about it, were filled and infested with it, and that to such a degree as men could hardly discern one another from the cloud."

The problem that Evelyn described did not get better. It got worse. During the Industrial Revolution, London's choking fogs grew increasingly dangerous. In 1905, Harold Des Voeux, an English physician, formed the word *smog* out of a rear-end collision between *smoke* and *fog.* A quarter of a century later, he founded the Coal Smoke Abatement Society. Des Voeux was fighting an uphill battle.

As usual, it took a disaster to raise public awareness of the problem and to galvanize people into action. In 1952, a high-pressure system smothered London in smog. This was not a picturesque Sherlock Holmes sort of fog. It was killer smog, and four thousand people lost their lives. Four years later the British government enacted the Clean Air Bill, which prohibited the combustion of soft coal in London. The result, almost immediately, was much cleaner air. Clearly the dangers of air pollution became an issue, at least one that legislators were willing to tackle—in opposition, as usual, to myopic business interests—only because it was attracting attention by killing too many people at once. (In China Miéville's 2007 fantasy

Un Lun Dun, the smog banished from Britain has become an evil plague in the parallel world of Un-London, where it feeds upon smoke and other pollution and grows ever stronger.) In the United States, the turning point was the 1970 Clean Air Act. Nowadays no reputable scientist refutes the data: elevated levels of particulate matter in the atmosphere translate into more deaths and a rise in health problems for those who survive.

※

Air pollution exists in several forms, each with its distinctive sources and health consequences. To make the situation more complicated, the effects of pollutants are often felt far beyond the source of the emissions. Winds blow in the real world, not on maps. They ignore political boundaries. But they affect us every day, in many invisible ways, and our survey of the day would be remiss without a glance at them.

In the United States, the Environmental Protection Agency defines toxic air pollutants as those that are "known to cause or are suspected of causing cancer, adverse reproductive, developmental, and central nervous system effects, and other serious health problems." The tedium involved in tracking real pollutants across imaginary lines results in a never-ending headache for scientists and for policymakers. For example, Toxaphene, a pesticide once popular in the southern United States, has been found in the Arctic, concentrated in the fatty tissues of polar bears.

To explain the herculean challenge of tackling air pollution, scientists emphasize the variety of its causes. Although other chemicals are known to be potentially dangerous, the EPA regulates almost two hundred pollutants as hazardous. They range from heavy metals, including lead and mercury, through deliberately manufactured chemical compounds, including pesticides and PCBs, to industrial by-products such as dioxin. An example will help demonstrate the way these toxins work. Nature does its part in producing atmospheric mercury through volcanoes, geysers, and even forest fires. But far greater amounts get pumped into the atmosphere by coal-fired power plants, incinerators that burn medical and municipal waste, and the type of cement kilns that burn hazardous waste. Manufacturers use mercury in barometers, light switches, batteries, paints, and, of course, thermometers. Unfortunately, prolonged exposure to mercury, either through swallowing or inhaling it, can damage the central nervous system of humans and other animals. The Mad Hatter in Alice's Wonderland was mad simply because he was a hatter. For generations, hatmakers used mercury to treat the fabric used in their work.

And mercury is only one example; each toxin behaves differently. Some break down quickly and have a health impact only in their immediate surroundings, while many others, like mercury, persist in the atmosphere and accumulate in intensity. But we seldom experience air pollution as the impact of a particular pollutant. If we did, it would simplify both scientific measurement and the legislation that results. Instead, we experience combinations and interactions of pollutants. When air pollution is discussed on the news, the terms you hear most often are *particulate matter, ozone,* and *acid rain,* with additional references to the natural phenomenon of pollen. Each of these problems requires its own studies and its own monitoring standards, and they gang up to cause more trouble than any one of them could generate on its own.

Particulate matter, sometimes referred to as *suspended particulates,* consists of those solid particles and liquid droplets that can be found floating in the air around us. The tiny particles scatter and absorb light, interfering with our ability to see objects at a distance and helping create the grim blur we accept as part of our urban lives. We describe it with the innocent term *haze.* Besides being a national issue of great concern, haze is also an imprecise but vivid index to levels of pollutants. In general, because of greater amounts of pollutants and also because of higher levels of humidity, haze is a worse problem in the eastern than in the western United States. According to the EPA, visibility in the eastern United States should be roughly 90 miles, but pollutants have reduced the range to 14 to 24 miles. In the West, visibility has shrunk from an average of 140 miles to 33 to 90 miles.

Particulate matter comes from a variety of sources. Nature throws spores, pollen, spray, and fog into the mix, but human contributions are more ominous. Some come directly from utility and industry smokestacks and automobiles, but others are formed by the reaction of chemical compounds in the air. Thanks to the variety of sources, the size of the particles varies greatly. More stringent standards for particulate matter now monitor particles that are 2.5 microns or less in diameter, which is roughly one-seventh the width of a human hair. In the past, the monitored particles were no smaller than 10 microns. The reason for the new standard is the growing body of scientific evidence that very fine particles pose a greater threat to human health, because they bypass the body's defenses and penetrate deeply into the lungs.

Nitrogen oxides are versatile villains, and they contribute to more than just ozone. They also unite with sulfur dioxide to form acid rain. Sulfur dioxide is a toxic compound primarily produced when power plants burn coal, oil, and natural gas. There are other sources too, but in the United States 70 percent of sulfur dioxide emissions come from power plants. *Acid rain* is the common name for what is more precisely called *acid precipitation,* because it can take the form of rain, snow, sleet, and even mist. (An even larger category, *acid deposition,* includes the fallout of dry particles.) Precipitation is considered to be "acid" when its pH balance is significantly lower than neutral. The pH scale represents the acidity or alkalinity of a solution. It ranges from 0 to 14, with neutral solutions at pH 7. Higher values are basic (or alkaline), with the scale going up through ocean water and baking soda to substances such as ammonia. Lower values are acidic, moving down the scale to vinegar (pH 3) and lemon juice (pH 1 to 2)—the ranges where acid rain shows up. The degree of acidity in acid rain varies, but "pure" rain, untainted by pollution, would have a natural pH of roughly 5.5.

Acid rain forms when sulfur dioxides and NO_x combine with water vapor to form sulfuric and nitric acids. While still in the air, these substances contribute to haze and other health-endangering effects. Then, when they unite and fall to the earth as precipitation, they damage soil and plants and make streams and lakes too acidic to support their natural variety of plants and animals. In a parallel phenomenon more visible to most of us, acid rain causes the deterioration of monuments, buildings, and even automobiles. Nor does it stop there. Acid rain can also pick up toxic metals as it filters down through soil, ultimately transporting them into streams and lakes—one way that such toxins accumulate in water and begin to climb up the food chain.

Although it hasn't yet been regulated by Congress, floating around out there in the air, along with NO_x and VOCs and the rest of the alphabet soup, is one of nature's least popular contributions to summer discomfort—pollen. It ranks high among the many natural and manufactured substances to which some people are allergic. Responses range from mild discomfort to crippling illness, and those unpleasant responses have inspired countless treatments. Pollen consists of grains or spores containing the male contribution to seed plants. It's produced by the anthers, and fertilization of the plant occurs when the pollen leaves the anthers and moves to the female pistils. Plants have evolved a number of ways of arranging this exchange of genetic information; sometimes it's even carried by insects or birds. But the vehicle that most affects human beings is this rising late-afternoon wind that is carrying pollen to your eyes, nose, and throat at this moment, and inspiring this survey of the hidden evils of our everyday air.

The Two Most Beautiful Words

Despite the perils of pollen and ozone and suspended particulates, most of us still manage to enjoy a hot summer afternoon. In temperate zones, the sun doesn't appear as our enemy unless we doze off on the beach and wake up with the outline of a swimsuit stenciled on our skin. We are more likely to revel in the fine pleasure of the overhead sun warm on our arms and neck and back, especially on those summer days when you can roll up your sleeves or wear a T-shirt to work in the yard and get sweaty and reddened by labor. Tolstoy's Levin exults in this primordial activity: "The sweat that drenched him cooled him off, and the sun, burning on his back, head and arm with its sleeve rolled to the elbow, gave him firmness and perseverance in his work." Engaged in such toil, you can't help but remember that when our revered ancestor *Australopithecus afarensis* was sprinting across the African savanna in pursuit of lunch, she was unaware that her descendants would spend their days operating a photocopier.

But even classroom- and office-chained urbanites flock outdoors on springtime lunch breaks and stand around simply basking in the welcome sunlight. During the warm months, who doesn't anxiously monitor weather forecasts as the weekend approaches? Despite warnings about skin cancer, we flock like lemmings to the edge of the sea and happily flop down to bake. Do other creatures experience pleasure when emerging from hibernation or even when venturing out from their burrows each morning? Surely woodchucks sit upright in their tidy doorways not just to see their shadow and predict the arrival of spring but also to enjoy the warmth of sunshine on their fur, like dogs and cats and the rabbit in Wallace Stevens's famous poem. A lazy and slow-moving afternoon outdoors, away from work, can be a time when you feel that with a few more days like this your real life might possibly catch up with the frantic fiction you suspect that you've been living. Nor would the purity of this experience be sullied by a nearby margarita and the Talking Heads singing, "Letting the days go by ..."

At this period of the day, there seems to be more time for thinking—or, if *thinking* is too dignified a term for this pensive observation, call it simply *gazing.* With the lowering sun at your back, shadow giants stretch ahead and parody your every step. Light poles and their penciled shadows triangulate distances across a parking lot. You notice for the first time, amid the sea of pavement surrounding shopping malls, that the shadows of the ornamental trees marooned on islets of bark have incredibly skinny cartoon trunks. As the sun sinks, the little trees' round cloud of shade lies on the pavement at the end of a thin shadow trunk, quiet and still, far from the mass of restless leaves that cast it.

In her autobiography, Edith Wharton—with a little help from Henry James—distilled the glories of a languid afternoon into one scene. In the happy days before World War I, they were visiting the medieval splendor of Bodiam Castle in East Sussex. In the seemingly windless sky above, peaceful white clouds hung still above a silent landscape of crumbled ruins. James and Wharton sat quietly admiring the buzz of dragonflies and the spots of color from flowers in the moat. Neither knew what the other was thinking, but the first remark was not about British history. "For a long time," remembered Wharton, decades after James died, "no one spoke; then James turned to me and said solemnly: 'Summer afternoon—summer afternoon; to me those have always been the two most beautiful words in the English language.' "

And does a summer afternoon offer a simpler or more portable luxury than aiming your face upward like a sundial to feel the sun's warm caress on your closed eyelids? The image of sundials reminds us that as our planet turns away from the sun we are nearing brillig, that time of day when the slithy tove, *Meles helicosaurus,* is likeliest to gyre and gimble. If you look above the wabe at the sundial's upward-gazing face, you will find that the shadow of its nose rests, as Humpty Dumpty explains in defining *brillig,* at "four o'clock in the afternoon—the time when you begin *broiling* things for dinner." Although *Through the Looking-Glass* was not pub-lished until more than a decade later, in an 1855 pamphlet for his family Charles Dodgson presented the first verse of "Jabberwocky" and defined *bryllyg* less narrowly as "the close of the afternoon."

The Silent Concert

There was a tiny blink of sun peeping in from the great street round the corner, and the smoky sparrows hopped over it and back again, brightening as they passed: or bathed in it, like a stream, and became glorified sparrows, unconnected with chimneys.

—Charles Dickens, *Dombey and Son*

Glorified Sparrows

Some photographers define "sweet light" as whatever lighting perfectly suits the mood they're after, whether it backlights sea oats at dawn or sidelights a formal studio portrait. But many professionals use the term to mean the warm golden light available very briefly just after dawn or just before sunset. The most overlooked ordinary scenes—a rusty basketball backboard, a sand-buried hurricane fence, a bedraggled starling on a dirty street—suddenly gain new significance, merely because they seem abruptly to be touched with grandeur from the sun's first or last light. Unusual weather conditions, such as the yellow sky around a tornado or a sudden break in an afternoon thunderstorm, can create similarly rich and dramatic lighting. Sweet light is also flattering to the entire range of human skin tones, which is one good reason why a first date ought to be outdoors in warm weather: the classic sunset dinner and moonlit stroll on the beach. As Dickens realized, special conditions can create dramatic lighting at any time of day, light that will glorify sparrows or anything else it touches.

> *When the sun begins to go down, its reflection takes form on the sea: from the horizon a dazzling patch extends all the way to the shore, composed of countless swaying glints.... "This is a special homage that the sun pays to me personally," Mr. Palomar is tempted to think.*
>
> —Italo Calvino, *Mr. Palomar*

The Path of the Light

Teatime and brillig are past. As the afternoon closes, and before day and night mingle in dusk, we are given the spectacle of sunset. The piled Michelangelo cumulus that sailed the afternoon are now awash in pinks and purples. The cloud shapes are limitlessly varied—dustballs and the prows of yachts, Winslow Homer's waves at sea, zeppelins and halves of mythic animals. All of them are spotlit with unexpected hues and angles of light. Along comes a high flotilla of white clouds, as precisely shaped if not as uniform as those in Georgia O'Keeffe's mural, and as you watch they grow ominously darker, the white retreating to a glowing shoreline on dark islands. One vast titanic cloud bank sails leisurely by. For a few moments the blue-gray clouds are underlit by a lower sun—a yellow-gray highlight that reveals more valleys and summits on this upended mountainscape than you could ever have imagined before the light moved. Some clouds look as substantial as the darkening earth below—formerly a vista of cooking fires and caravans, now shadow-chilled abysses between office towers whose upper stories still dazzle with the last of the sunlight.

Suddenly a jet seems to burst into fire, but then you realize that the sun has merely noticed it. Its drifting contrail becomes a gash that reveals a flaming cosmos ordinarily hidden behind the strong blue bowl of the sky—Heraclitus's fire at the heart of creation. The light, it occurs to you, was already up there, glancing through the atmosphere, scheduled to arrive exhausted at a seaside restaurant in Cozumel, the photons lingering just long enough to titillate the pixels in a *turista*'s camera. But the photons were intercepted. This airplane's curved metallic shell deflected them like a defense weapon and bounced them down into your upward-gazing eyes. The thought of your own eyes pulls your consciousness outside yourself and makes you realize that the reflections of the sunset must be parading across your wet corneas. To see the heart of creation, you need only stand in the path of the light.

※

Over the last few millennia, humanity's own brief moment in the sun, our responses to this colorful show every afternoon have been as varied as one could imagine. "The sun was now gone out of the sky," writes Hawthorne in one of his stories, "leaving, however, a rich inheritance of his brightness among those purple and golden clouds which make the sunsets of winter so magnificent." The aggressive color can inspire less peaceful imagery than a rich inheritance. We have been mortal animals, both predator and prey, for too long to stop associating red with blood. The sad narrator of Juan Rulfo's novel *Pedro Páramo* reads violent symbolism into the sky over a heartbreaking departure: "The day you went away I knew I would never see you again. You were stained red by the late afternoon sun, by the dusk filling the sky with blood." And the narrator in Arno Schmidt's *Scenes from the Life of a Faun*—an angry, depressed man bitterly witnessing the rise of the Third Reich—sees the clouds themselves as sanguinary: "*At the very last:* the bloody torsos of clouds piled up in the west, the light's mass grave, drenched in gore, in smoke."

※

Some writers have a less violent but equally personal approach to this daily spectacle. "Today," confides one man to his diary, "I saw a red-and-yellow sunset and thought, How insignificant I am!" Before we nod accord, you should know that the perpetrator of this discount existentialism is Woody Allen, and that the narrator of his story immediately adds, "Of course, I thought that yesterday, too, and it rained." The moment that we try to enjoy this melodramatic time of day we find clichés scribbled across it like graffiti. Sunsets can be senior-prom melancholy and often too cinematic to be real, but most of us are susceptible to cheap special effects. Many people sleep through sunrise, but almost everyone who doesn't work deep inside a windowless mall or tending an underground missile silo at least glimpses sunset. This time of day's bright colors, its picturesque silhouetting of trees and bridges, its flattering light for close-ups, its easy autumnal symbolism—all have been extolled by promotional calendars and bad romantic movies, by TV commercials for insurance and laxatives, until a natural daily occurrence is in danger of becoming self-conscious, ashamed to appear showy and gauche.

Andy Warhol, who specialized in being showy and gauche, once filmed an entire sunset, with no action visible for several minutes except a sinking sun and a jet drawing a contrail across the sky. The silent, washed-out little movie doesn't live up to the grandeur of its subject. This is partially because the film stock lacks the ability to record more than a hint of the subtle tones, but also because Warhol seemed to be deliberately reducing an astronomical event to the banality of a cheap home movie—or perhaps he was merely boiling the Hollywood out of the experience. Like Warhol's cardboard boxes of leftover trash, his pallid film of the sunset resides in the archives of the Warhol Museum in Pittsburgh, guarded by worshipful attendants. Outside the museum the sun is setting. Someone turns on a light.

*

Let us return for a moment to Milo, the adventurous hero of Norton Juster's children's book *The Phantom Tollbooth*. When last we saw Milo, at midday today, the orchestra conductor Chroma the Great was showing him how the world might look without color. Chroma conducts the silent thousand-piece orchestra whose daily performances create the ever-changing colors of the day. As Juster realized, the concert of the sun requires many players. His personification of the sun's daily circuit of the sky is a charming fable for children; it needs no other justification. And, like the rest of the book, it contributes to the author's unimpeachable "moral," the story's cumulative goal of promoting attention to the mysterious realities of everyday life. The story of Chroma's orchestra is relevant to our journey through the day, however, not just because it is about daily cycles but because it works precisely like mythology.

Juster nicely envisions the instrumentation when Milo witnesses the concert of the sunset. He has the musicians fade out their instruments, one at a time, as the color seeps from the sky. Finally only bass fiddles are continuing to sound the night, and a set of silver bells "brightened the constellations."

Milo tells the conductor that his sunset was very beautiful.

"It should be," replies Chroma; "we've been practicing it since the world began."

Diamond Dust and Sun Dogs

Like real light, symbolic light refracts differently in each medium; sometimes we find it glimmering far from its source. In Christian iconography, for example, a saint usually wears a halo. It's a shame that this insignia gets issued only posthumously, like a war medal; think how helpful it would be in real life. In art it also adorns other key figures—god, demigod, angel. The traditional Christian catalog, largely from the Middle Ages and Renaissance, offers several flattering styles of halo: a broad ring around the planetary head, rather like one of Saturn's; a flat plate like the brim of a shiny hat; a tiny wirelike ring like the orbit of a mosquito; or the ever-popular semicircle of gold filigreed with punch-hole patterns. Not that virtuous effulgence is always so precisely drawn. Presumably the slapdash glow behind the throne of Michelangelo's adjudicating Christ on the Sistine ceiling represents a halo, although it looks more like the light of an approaching train.

What all these haloes have in common is a borrowing of glory from the sun and moon. Haloes are not entirely imaginary. Meteorologists use the term *halo* to describe the rings that you can see occasionally around the sun or moon. These are caused by our looking up through either a cloud composed of ice crystals, such as cirrostratus, or through *diamond dust.* This evocative term describes minute crystals of ice—unbranched, not feathery like snow crystals—that hang in the air or float gradually to the ground, precipitating from high clouds or even from a perfectly clear sky. They refract light. Sometimes the rings—or arcs, if only part of the ring is visible—are whitish and sometimes colored. If the latter, you will see blue or violet on the outer edge and red on the inner. The order of ring color reverses when a solar or lunar halo is caused by water droplets rather than ice. This kind of ring is called a *corona,* from the Latin word for "crown." The water droplets in a fine layer of low cloud, such as altostratus, diffract light from the sun or the secondhand sunlight bouncing down to us from the moon.

Right now, at sunset, a lucky skywatcher can sometimes witness an optical phenomenon called a *sun dog* or *mock sun*. "Dazzle mine eyes, or do I see three suns?" exclaims Edward in *Henry VI, Part III*, and he can't resist adding, "In this the heaven figures some event." On Shakespeare's stage anything can happen, but what actually goes on in the true global theater is a relatively rare but now well-understood optical phenomenon whose theatrical nature doesn't seem to require divine intervention.

Ice crystals come in a variety of shapes, but it takes only two to form most of the phenomena we glimpse up there in the omen-filled sky. The two most common forms are both hexagonal, providing them (counting the ends) with eight refractive surfaces. One crystal is a flat waferlike shape and the other a six-sided column like a pencil. A hexagon tends to refract light at an angle of 22 degrees, and if the high cold air contains enough columnar shapes, the result will be a halo around the sun—or around the moon, because the same effect works at night.

As ice crystals form at high altitudes, gravity orients them horizontally. The columns that form haloes fall as if turned over on their sides, but the plates fall flat, like flying saucers. The plates can also refract light at 22 degrees, like the columns, but they only do so when they have fallen to an angle above the earth that is equal to the sun's. Under these specialized circumstances, we are treated to the spectacle of sun dogs—brilliant spots of colorful light on each side of the sun, 22 degrees away from it, as it sinks into the mythical west.

A different momentary light show sometimes appears directly above the setting sun. In the same way that refraction can make the moon look flattened at the horizon, so can refraction cause the phenomenon known as the *green flash.* You may have noticed that although the entire setting sun is orange, the area closest to the horizon seems darker, red-orange or red. This is because the atmosphere's prism is separating the light into its various colors. So it is not surprising that when conditions are just right, you may glimpse, for a second or two, a flattened oval flash of green or blue just above the setting sun, with the rest of the spectrum below it: yellow, then orange, then red at the horizon.

Occasionally at this time of day you can also observe what is called a twilight rainbow. As can happen at sunrise, sometimes at sunset or just afterward a rainbowlike arc forms that is almost entirely red, the other colors having been filtered out of the light rays as they angle through the atmosphere. After Newton deconstructed the spectrum in the late seventeenth century, natural philosophers paid a great deal of attention to rainbows, especially to variations in the natural projection of a spectrum—the white arcs occasionally produced in mist, the colored haloes around the moon, and the rare but striking appearance of the twilight rainbow.

Sunset Races

Go west, young man.

Westward, ho!

The Wild West.

The West of the American imagination, from the land beyond the first coastal colonies to the genocide that accompanied dark fantasies of Manifest Destiny, is the land of the setting sun. Henry David Thoreau, predictably finding every aspect of life metaphorical, and automatically writing from the point of view of a white descendant of Europeans, held forth in an essay: "We go eastward to realize history and study the works of art and literature, retracing the steps of the race; we go westward as into the future, with a spirit of enterprise and adventure. . . . Every sunset which I witness inspires me with the desire to go to a West as distant and as fair as that into which the sun goes down. He appears to migrate westward daily, and tempt us to follow him. He is the Great Western Pioneer whom the nations follow." And he added, forming a couplet half from a quotation, "To Americans I hardly need to say,—

'Westward the star of empire takes its way.' "

The West, the home of the setting sun, has always inspired stories; they just varied in their implications from place to place. In the *Odyssey*, the island king Alkínoös introduces Odysseus—who has been cosmetically modified by the doting Athena, as usual, to appear taller and broader and more of a movie star—as a guest who arrives "after long wandering / in

Dawn lands, or among the Sunset races." Elsewhere in the same story, Menalaos tells Telemachus that his father traveled to Egypt "and still farther among the sun-burnt races." How handy that, from the perspective of Rome, northern Africans lived in the land of the sunset. A millennium later, Pliny declared with his usual certainty that there is "no doubt that the Ethiopians are burnt by the heat of the sun and are born with a burnt appearance and with curly beards and hair." Then he added what for him passes as evidence: "In the opposite region of the world the races have white, frosty skins with flaxen-coloured hair that hangs straight." Whether Pliny considered certain incidents from the epic tragedy of Phaethon as evidence for his skin color theory is not known, but he was familiar with how the themes intertwined, as in the following section about the reckless chariot ride of Apollo's son.

The Nile Dries Up

Although Phaethon knows it is too late now, as he becomes a passenger aboard a runaway chariot in the sky, he repents his hubris, his arrogance, his entire journey to Apollo's home. He wishes he had never gone to his mother and demanded proof of his father's identity. The guilt and remorse are as dizzying as their speed above the earth. Paralyzed, Phaethon looks back at the far horizon of the earth and day behind him to the east, then turns to see an equal distance of sky ahead to the west. Should he keep going or return? Is he even able to influence these wild-eyed horses? He cups the reins without guiding the horses and realizes in despair that he doesn't even know the animals' names.

But far more terrifying than the beasts who draw his chariot are the giant mythic animals that Phaethon sees before him in the sky. For he is high enough now to not only startle the constellations but to crowd them. Ahead lies the undulant form of Scorpio: two curved pincer claws, the much broader curl of the tail spanning the width of two lesser constellations. Seeing the reckless chariot and wheezing horses, the Scorpion arches the bow of itself, its tail like a venomous dart. Staring, paralyzed, Phaethon drops the reins and they slide across the horses' frothing backs.

The horses feel the slack reins and lose all restraint, tearing from Apollo's familiar track and plunging into uncharted sky. Avoiding the Scorpion, they stumble into stars and scorch the clouds. The moon, lingering pale in the west as if reluctant to abdicate her own throne, is dumbfounded to watch her steadfast brother's steeds running lower than her own. The chariot careens toward the sky above and then pitches, rolls, and plummets closer to the earth.

The chariot of the sun is coming too close and the world is burning. The higher altitudes begin to brown, smoke, then burst into flame. Moisture evaporates from the soil as hissing steam, and even rivers exhale their boiling waters into the air. Whole hillsides of trees catch fire, fresh leaves below feeding the flames that start above. Even the sacred sites are not safe. Fabled mountains begin to burn: Etna, Helicon, the Alps and Apennines. City walls tumble as fire sweeps across them and people race away screaming, some only to fall under timbers or catch fire themselves. Gripping the edge of the chariot with white knuckles, his hands free of reins, Phaethon gazes down in horror and feels a heat like a blacksmith's furnace burn his face. Smoke clouds his view and soot flies into his eyes.

The seven mouths of the Nile dry up as it flees to the edge of Earth. Great cracks form in the parched surface, permitting light to shine down into the underworld, past Avernus, the crater that leads to the realm of the dead—where the rulers of eternal night look up, painfully squinting in consternation. New islands jut from the sinking ocean. Now is the time, the Greek and Roman stories will say later, when Libya lost its fabled springs and turned to desert, and blood rushed to the skin of Ethiopians and turned them dark.

All Those Cares and Fears

A sunset doesn't have to result from divine irresponsibility to reflect the network of causes and effects that color the sky above us. Nor are all sunsets quite so lyrical as Norton Juster's version, or at least they may seem less romantic when you think about what lies behind some of them in place of Chroma's artistry.

You may have noticed that sunsets vary in intensity not only randomly, depending upon a certain area's accumulation of dust or a certain day's allotment of clouds, but also in response to other natural phenomena. Sunsets are an informal measure of the amount of dust and other particles in the air, which is one reason why the close of a winter day is often less dramatic than sunset in warmer months.

Often sunsets reflect major atmospheric turbulence far from its source. In 1991, for example, after more than six hundred years of dormancy, the Philippine Islands volcano Mount Pinatubo erupted. Because it was so carefully watched by scientists—many of whom barely escaped with their lives from the American naval base nearby—Pinatubo has become a test case for atmospheric disaster. The eruption column, with an 18-kilometer-wide base, rose 30 kilometers into the air, ejecting more than five billion cubic meters of ash and other particles into Earth's atmosphere. The estimated 15 to 20 million tons of sulfur dioxide that came pouring into the air joined with water to double into between 30 and 40 million tons of sulfuric acid particles. One result, not surprisingly, was dramatic sunsets over faraway parts of the world for the next couple of years. Scientists think that another side effect was the global atmospheric cooling that occurred during the next year. Middle and lower atmospheric temperature tests fell almost 1 degree Fahrenheit—and, some atmospheric scientists remarked, might well have been lower if not for the rise in Pacific Ocean temperatures because of El Niño. As you may recall, however, the sunsets afterward were magnificent.

Dramatic sunsets also accompanied another famous atmospheric upset. In 1812, the volcano Soufrière erupted on the West Indian island of St. Vincent; two years later Mayon erupted in the Philippines. Then, in 1815, a volcano named

Tambora erupted with fifty times the force of Mount St. Helens, blowing almost a mile off its height and darkening the sky for more than three hundred miles. The force of the eruption carried fine ash particles into the upper atmosphere, where they formed an almost global cloud that intercepted sunlight. Sunsets all over the Northern Hemisphere were theatrically colorful, and again proved an index to further trouble. The next year, 1816, is popularly known as "the Year without a Summer." Apparently the drop in solar radiation reaching the surface didn't lower winter temperatures dramatically, but the winter temperatures stayed through the summer. Instead of rising temperatures in the spring and later, Europe and Canada and the United States experienced an often subfreezing replacement for the normal summer. Countless crops failed and the result was widespread famine. Post-Napoleonic France may have been hit the hardest, with very ill-timed starvation fomenting more civil unrest.

But history's hindsight can see a glimmer of positive return in even these sad records. By June of 1816, many parts of Europe were going through long periods of steady rain. Such was the weather in Switzerland, on Lake Geneva, as four English tourists abandoned their boating plans and huddled indoors around a fire. They were the poet Lord Byron, the poet Percy Bysshe Shelley, Shelley's brilliant young mistress Mary Wollstonecraft Godwin, and Byron's physician and sycophant, John Polidori. The gloomy weather led the bored quartet to launch a ghost-story contest, which the young woman unquestionably won with *Frankenstein*, a novel that now bears her married name, Mary Shelley.

And naturally young Victor Frankenstein was an easy target for a moody sunset:

> The wind, which had hitherto carried us along with amazing rapidity, sank at sunset to a light breeze; the soft air just ruffled the water and caused a pleasant motion among the trees as we approached the shore, from which it wafted the most delightful scent of flowers and hay. The sun sank beneath the horizon as we landed, and as I touched the shore I felt those cares and fears revive which soon were to clasp me and cling to me for ever.

Sic Transit

Worldwide cooling in response to atmospheric trauma—the kind that occurred after the eruption of both Tambora and Pinatubo—inspired the scenario behind a scientific interpretation of the past on Earth and an extrapolation about an alternative future. Both are frightening to contemplate as we gaze at the sunset today. After the colossal impact of the meteorite or asteroid that struck Earth about 65 million years ago, in the area of what is now the Yucatán Peninsula, fires burned much of the surface, and then a global cloud of thick dust blocked sunlight and prohibited photosynthesis. The result was the extinction of three-quarters of the animal species on Earth, including (most famously) the dinosaurs. Millions of years later, Carl Sagan and planetary scientist James Pollack were examining a curious global dust storm on Mars when spectrographic analysis indicated that temperatures were higher above the dust storm and lower below it than they would have predicted. Deciphering this atmospheric phenomenon, applying it to the far busier atmosphere of Earth, and mixing in knowledge of the extinction-causing asteroid led Sagan and other scientists to propose the "nuclear winter" scenario: the prediction that global atmospheric darkening and cooling, in the aftermath of a nuclear war, would cause further environmental disasters and drastically lower the temperature worldwide. It was an apocalyptic picture worthy of some dark religious vision.

✳

Even if we manage to avoid blowing ourselves up and blocking out the sunlight over our ruined world, there are plenty of other ways the world might end, and several involve the sun. Aztec mythology described five separate times in which the cosmos was created. The present time is the fifth. The earlier four, each of them associated with one of the four prime elements (earth, air, fire, water), ended with the sun itself being thrown from the sky. This is not a good omen for us.

Although it is not a mythological figure, the sun embodies some mythological paradoxes. It is not alive, but we say that it was born and will die: 16 billion years ago and perhaps 50 billion from now, by the latest ballpark estimates. In the nineteenth century, astronomers theorized that the sun would eventually grow cold and burn out like a candle. Nowadays a greater understanding of stars' cycles indicates that ours will go through a long, slow red-giant phase, expanding and then shrinking and finally sputtering away into darkness. No one will be on Earth to observe this phase because our planet will have long since been vaporized or at least reduced to a cinder, as surely as if the Aztec gods had thrown it from the sky.

At every moment, the sun is using up its own fire. For physical processes as for animal lives, living is dying. The sun's profligate expenditure of energy includes the expulsion of its own substance as an aura of solar wind—ions and electrons radiating outward at speeds of up to 500 miles per second. The magnificent fluctuating curtains of color in the aurora borealis (and its southern twin, the aurora australis) occur when charged particles of solar wind encounter the earth's magnetic field. The magnetosphere wards off most particles, but those that are captured spiral along otherwise invisible lines of magnetic field. Some of the energy produced during this bombardment is released as brilliant light, and it occurs near the poles because it is there that the earth's magnetic field emerges. Solar flares—sudden bursts of solar wind—cause the aurora to dance more wildly and even to interfere with electrical transmissions on earth. These are merely some of the many processes that are involved in the slow aging and death of a star like ours.

Larger stars burn out more quickly. The truly massive can collapse upon themselves and then explode into a supernova that, once it sheds the resulting shell of layered gases, may shrink into a neutron star no larger than a city on Earth but with a literally inconceivable density. The largest stars can mutate ultimately into the smallest and densest of astronomical phenomena, a black hole. In this violent universe, it seems less disturbing to imagine our own star fading quietly into senility.

And yet, that the sun will die at all is disquieting enough for some people. In a letter to fellow scientist Joseph Hooker, Charles Darwin expressed his vision of the future, imagining natural selection as a kind of biological determinism leading inevitably toward social and anatomical progress. Then he added,

> I quite agree how humiliating the slow progress of man is, but everyone has his own pet horror, and this slow progress or even personal annihilation sinks in my mind to insignificance compared with the idea or rather I presume certainty of the sun some day cooling and we all freezing. To think of the progress of millions of years, with every continent swarming with good and enlightened men, all ending in this, and with probably no fresh start until our planetary system has again converted into red-hot gas. *Sic transit gloria mundi* with a vengeance.

Many astronomers and many writers have tried to imagine these last days of our star, with wildly varying attention to the likely actualities. The *Dying Earth* cycle of stories by the American fantasist Jack Vance is set in a period when the sun is growing old, which Vance once described this way: "The time is the remote future. The sun gutters like a candle in the wind. . . . In ruins along the coasts of Ascolais and Almery a few languid men and women amuse themselves until that hour when the sun finally glimmers into darkness and the earth grows cold."

Feeding Time at Loch Ness

Twilight was already gathering in the luxuriantly leafy trees of the long road, and the clouded sky had paled to a sort of lightless hazy white.... The lazy capacious brick houses had put their lights on and uncurtained windows cast squares of brightness upon neatly raked gravel or attentive cascades of rambler roses.

—Iris Murdoch, *The Sacred and Profane Love Machine*

The Evening and the Morning

We use the word *daytime* to mean the hours lighted by the sun, but we use *day* to refer to the full twenty-four-hour period of Earth's rotation on its axis. As a consequence, at twilight we must ask ourselves a seemingly paradoxical question: does the arrival of nighttime launch the night or the day? The answer seems obvious only if you don't stop to think about it. When does a day begin? Why then and not some other time?

Despite the clocks that measure hours before we awaken, many people still tend to think of each day as beginning in the morning, usually at the time when we began our journey today—at dawn, just before sunrise. But this moment of demarcation is only one of many alternatives. Yes, the ancient Egyptians began measuring a new day as we do, at sunrise, but what about the ancient Hebrews? They began each new day at sunset. The order of the lovely phrasing in the Genesis creation myth is telling: "And the evening and the morning were the first day." And here we are in the twenty-first century, thousands of years later, and Jewish holidays still begin in the evening and extend through the next day until sunset. No wonder traditions can be so binding. They grow and adapt like living beings, retaining enough of their original form to be recognizable, but changing enough to survive in a new (and frequently hostile) environment.

Christianity, for example, started out as a rabble-rousing faction of Judaism, so it isn't surprising to find remnants of these ancient Hebrew ideas in contemporary Christian observances, especially in the tradition-festooned halls of the Catholic Church. When Christianity moved away from its theological birthplace, it packed a bagful of Hebrew traditions, but many of those—such as the idea for a seven-day week, with the last day for resting—the Hebrews had already inherited from Babylon during the dark years of captivity.

Consider the "eve" that precedes so many Christian holidays. All Hallows' Eve (Hallowe'en, from Hallowed Evening) is the night before November first, which is All Saints' Day in Catholicism. This formulation lingers on because the holiday originally began in the evening, then broke off to form a separate eve when traditions began to change. The same applies to Christmas Eve and New Year's Eve. We celebrate the full day of the holiday, but add a sort of half-holiday from the night before. (This is worth remembering in case you yearn to open your gifts a few hours early.)

Then we have the question of how to define morning. In our clock-obsessed world, we don't start counting hours when the sun comes up. We start a new day in the middle of the night. Why? In the medieval church, daytime and night were each divided up into twelve hours, a remnant of the rare duodecimal (base 12) counting system. (Incidentally, minutes and hours are from the sexagesimal—base 60—counting system that was itself borrowed from the Babylonians, who got it from the Sumerians.) The twelve hours of daytime began at sunrise, the twelve of night at sunset. In a society without subway schedules and *TV Guide*, the length of the hours mattered less than their assigned numbers, so their duration varied depending upon the season. A night in winter had longer hours than the same period in summer, but there were still only twelve.

This method sounds odd to us, and yet it persists in the church's *Liturgia horarum,* the Liturgy of the Hours—better known as the Canonical Hours. The prayers for these times reside in a book of hours, which is basically a lay version of the breviary carried by priests. The ritual of daily prayer repeated at specified intervals was itself well established in Hebrew observances long before Christianity. Many religions encourage the use of some kind of physical reminder or counter of daily prayer. Tibetan Buddhists employ a prayer wheel, the *mani choskhor;* Muslims and Catholics count prayer beads, the *subha* and rosary respectively. In Catholicism the rosary was originally the recitation of prayers, which finally lent its name to the beads that tracked the recitation. The Canonical Hours are intended to consecrate the entire Christian day by providing at three-hour intervals a spiritual milestone, like a rosary knot that focuses contemplation. Some commentators have described the Hours as divine oases in the desert of the godless day, just as the sabbath and other holidays (holy days) frame and enliven and sanctify the earthly progression of the secular week.

Even those of us who don't find the godless day particularly arid can still find interest in the Canonical Hours. In their naming and timing, we find the fossilized genesis of the twelve-hour day and twelve-hour night. Twelve midnight, the starting point for our current definition of the day, is the canonical hour *sext* (six) because it was originally counted from the evening, just as 6 a.m. is *prime* (one) and 9 a.m. *tierce* (three). Three p.m. is *none* (nine), a word that migrated backward and became the name of our midday—*noon.*

☀

Vespers, the sixth Canonical Hour, stands out as a nature-born word amid the numerals. It comes from the Latin *vesper,* via Middle English, and it refers to a phenomenon that you will be able to observe glimmering just above the horizon now if you look to the west: the evening star. This light in the sky stands out for two reasons—its magnitude and its timing. It is so bright not because it is a huge star or a near star but because it is not a star at all. It is the planet Venus. Because Venus is closer to the sun than Earth is, it appears to us to remain in the sun's neighborhood in the sky.

Naturally, no other light could compete with the star of the show, so Venus is invisible unless the sun is offstage. We see it shortly before the sun arrives or after it departs, which means that in both circumstances it seems to be the brightest star in the sky. Our closest neighbor, Venus is a twin of Earth in size but cloaked in highly reflective clouds that lend it a disproportionate magnitude. Probably Venusians don't consider dimmer Earth their evening star.

The trail of the Latin word *vesper* goes back to an ancient Indo-European word or prefix, *wespero-,* which referred to evening or night. One of Venus's many nicknames, Hesperus, derives from the Greek word *hesperos,* which also evolved from *wespero-*. The same fertile stock reproduced in other directions and eventually, via the Germanic word *westo-,* became the word that refers to the direction in which we are now gazing, the direction that fascinated Thoreau and Ulysses: *west.*

Uati, the Egyptians named this planet: first star of night. When they saw the same light lingering in the eastern sky before sunrise, they called it Tiû-nûtiri. In a fine example of the human inclination to think in analogies, the Egyptians also called Venus Benin, meaning "heron," because herons dip below the surface of the Nile and reappear, just as Venus disappears for a while but always returns. The celestial heron flies high above the gathering darkness, star of both the evening and the morning.

Two Lights

Above the western horizon, where the sun just set with such majesty, clouds are now invisibly the color of the sky except for scumbled pink sidelighting. In this low light, pale blossoms on shrubs and trees linger on the edge of visibility, romantically luminous as if with extra wattage. Behind them leaves are still rich green up close, already dark at any distance or against the pre-nighttime sky. Leaves at the edge of branches begin to lose the species-specific silhouettes that they acquired at dawn; they surrender individuality and sink into the anonymous mass.

In cities not all trees suffer this change of identity. Every urban dusk provides the Magritte vision of trees lit from within by streetlights—false day within night. Above them the sky is still too bright for all but the most assertive stars, yet not so bright that the already hidden sun can't set afire jet contrails. In a sunset breeze leaves seem as restless as the creatures hidden among them, or as the human mind observing them. Animals too forfeit identity and become scurrying shadows. Perhaps this phenomenon inspired shapeshifter myths—a shadow seems to be a rabbit but turns out to be a cat. Such changes take place constantly at twilight.

When you sit outdoors reading in the late afternoon, the light fades so gradually that your eyes respond without your having to consciously notice what is happening. The adjustment between the two parts of the eye occurs slowly, and twilight has become almost nighttime before you are completely unable to see the words on the page. But after a while you realize that, even though your eyes are efficiently processing the fewer and fewer available photons, the color is fading out of the world around you.

What is occurring is a switchover in the eye itself, from color-sensitive daytime vision machinery to a nighttime system that works better in low light. During the bright light of day your eyes have been using mostly their cone cells. These sensitive and efficient little receptors are common to almost all vertebrates except those that live permanently in dim light. They provide us with the ability to perceive color. (The quantity and sensitivity of cones in animals' eyes vary enormously. Cats and dogs, for example, each possess some; although their perception of color is quite limited, they are not truly color blind.) The retina in each human eye contains about six million cones, each with a supply of a pigment called retinene. The cones come in three styles. Each contains its own genetically determined protein, a form of rhodopsin, one of the several opsins that—along with serotonin, melatonin, and other substances—maintain circadian rhythms by responding to changing light levels. Each opsin responds to a different wavelength of light, enabling us to see either red, blue, or green, or variations thereof. Nonscientists call rhodopsin *visual purple,* an appropriately evocative term to be associated with twilight.

Now, at dusk, the much lower light level would result in near blindness if our eyes were not supplied with a backup system for nighttime vision. Your primate retina contains about 120 million rods. These highly sensitive secondary receptors respond to much less light than cone cells require. In providing us with low-light vision, however, there is a trade-off: rod cells barely perceive color at all, but are most sensitive in the blue end of the spectrum, unlike the yellow and red bias of the cone cells. Thus the black-and-white-movie aspect of nighttime, as well as the wonderful way that color seems to gradually pour back into the world at dawn and slowly seep out of it at twilight.

Twilight is a romantic word, like *dusk* and *sunset* and *nightfall*—words dreamy and pensive in a way that *midday* could never be. Do we experience this emotion in response to the lure of Dionysian recreation after the Apollonian work-day, the promise of lessened strictures and more exciting possibilities away from the sun's inhibiting gaze? Or do our associations go much further back in time than such recent human context? The daily reduction in light as the sun went down caused perceptual problems for animals long before human beings in cars were blinking their headlights at other drivers to warn them to turn on their own.

The end of the day—yet not quite nighttime—can be powerfully symbolic. Some of the darkest stories in mythology are Norse accounts of the Twilight of the Gods. The eighth-century Chinese poet Meng Hao-jan returned again and again in his poetry to images of twilight—delighting in its quiet mystery, declaring it the time when a traveler's loneliness returns. He captures this time of day's unique attributes when he writes about watching the figures of birds slowly fade away, "to fathom the erasure of insight." Dickens titled one of his quirky little ghost stories "To Be Read at Dusk." H. G. Wells's Time Traveller describes the Morlocks, when he first glimpses their scurrying pale forms at dawn or twilight, as "mere creatures of the half-light." And when John W. Campbell, the American science fiction writer and editor, wanted to write an elegiac story about a time traveler who returns to the present (of 1934) with a bleak vision of the end of humanity, naturally he titled it "Twilight," and naturally the story begins at dusk.

Consider the word for this misty borderland between daylight and darkness. *Twi-* is a combining form meaning simply "two," as in the Greek *di-* and the Latin *bi-*. *Twi-* itself comes from the Old English *twa* (preserved in, for example, Robert Burns's dialogue between talkative canines, "The Twa Dogs"). Add to it the Old English *leoht*, "light," and you have our word. Contemporary cognates include the German *Zwielicht*. It is interesting that *twilight* means "two lights" when this time of day experiences something closer to "half-light." Such ambiguity is apt. Are we experiencing the end of day or the beginning of night?

"Can you feel how all transitions," murmurs Rilke, "hesitate between one state and the next?"

Nessie and the Bat

It isn't just the air or light level that feels indecisive; even twilight's creatures appear undefined. In her autobiography Edith Wharton described a friend of hers, the reclusive man of letters William Brownell, as having a "shy and crepuscular" personality, "always an aloofness, an elusiveness." Her description nicely embodies the traits we associate with this time of day: indistinct and always changing. The word *crepuscular* derives from the Latin *crepusculum,* from *creper,* meaning dusky or dark. It refers not only to dusk's low light but to any creature active particularly at this time of day.

Animals adapt to the unique attributes of their surroundings, and borders are no exception. Twilight is a border habitat, like a hedgerow or Alsace-Lorraine. A hedge may harbor creatures adapted to lawn or street; Alsatians speak both French and German. Twilight is also the time of the changing of the guard. You may see animals that are active primarily in daylight or darkness—perhaps burrow-bound after a hard day preying on herbivores, or perhaps awaking for the second shift and looking for breakfast—but the unique character of this time of day will be most visible in crepuscular animals.

Countless animals are active throughout the day in different regions of the world, but we can't take time to examine all of them. The border where daylight and darkness mingle is inhabited partially by creatures of both, but more by animals adapted to its boundary characteristics. The best known is considered a border creature in every way. Almost bird, not quite mammal—this is how the bat shows up in folklore. In many places, however, the flight of bats at dusk is as characteristic as the birdsong chorus at dawn. They inhabit all but the coldest outposts of the globe, ranging in size from the sixteen-inch giant flying fox to the fingernail-size Kitti's hognosed bat.

Even in dense urban environments, you can usually glimpse bats foraging after the sun goes down. In low light their seemingly directionless flutter distinguishes them from birds, even from other fast-moving insectivores such as swifts and swallows. Like contestants in a quidditch game, they are in frantic pursuit—in their case, of insects that dodge on the wing. Bats are active much of the night, but only at twilight is it dark enough to encourage them to forage while still light enough to permit us to watch. Next time you are walking at dusk or attending an outdoor cocktail party, watch for them swooping above your head. They are also drawn to the insects that congregate around well-lit signs at night, especially those in the red end of the spectrum.

Leopold Bloom, at dusk on that immortal sixteenth of June in *Ulysses*, watches a scurry in the sky and wonders what he's seeing. Is it a swallow? No, probably a bat: "a little man in a cloak he is with tiny hands." Anyone who has seen a bat up close recognizes the description. Yet bats—despite their useful job as voracious insectivores—look less like a tiny man than like a tiny imp in a Bosch painting. In Christian iconography demons sport bat wings as reliably as angels wear dove wings. Bats have often been called the bird of the devil. In the metaphysical carnival that is his book-length poem *Notes from a Bottle Found on the Beach at Carmel*, Evan S. Connell has his narrator note at twilight that he mistakes the flight of a bat "for the passage of the Evil One." Far down the aesthetic spectrum, we find a fictional character who also admired the bat's dark symbolism, the superhero who seems an absolute antonym of Apollonian Superman—the Dark Knight of

Gotham City, Batman. Therefore it is worth remembering that Emily Dickinson, ever the appreciative contrarian, praised the elate bat as an example of God's beneficent eccentricities.

Surely the bat's broad appeal as a symbol derives from its seemingly ambivalent position among nature's twilight creatures. An African folktale recounts the shifting power balance between birds and beasts, which at last erupts into war. Gradually creatures of the air and creatures of the earth discover that the bat, hoping to side with whichever army triumphs, is professing loyalty to each. In time the land creatures win and they prosecute the bat for treachery. His defense attorney argues that indecision is part of the bat's nature; equipped with fur and teeth, he is a land animal, but wings also grant him access to the world of birds. In a similar fable, Aesop takes the idea of the bat's dual nature even further, explicitly linking it to twilight and using it to warn against the false security of neutrality. Lambasted by both sides, the bat takes to avoiding creatures of both night and day, never leaving his nest until after the daytime animals have retired—and even then cringing in corners and eaves. The same sort of unease in the presence of the indefinable has labeled frogs and toads, which are unwilling to commit to either a terrestrial or aquatic lifestyle, as suspiciously nonconformist; the *amphi* in *amphibian* (as in *amphitheater* and *ambidextrous*) means "on both sides."

※

Twilight also seems to be the feeding time of another creature relevant to our story—the Loch Ness Monster.

Apparently many people still believe in Nessie—despite murky and shamelessly doctored photographs, despite identifications and estimations of speed and distance by people utterly unqualified to make such guesses, despite unequivocal confessions of fakery in the name of tourist-baiting. The following point will demonstrate why scientists label such claims as beliefs rather than as hypotheses: one "researcher" actually insisted that the high percentage of Nessie sightings at twilight proves that the creature is either crepuscular or nocturnal. Never mind that this murky time of day plays tricks on our eyes. Forget that this in-between period is the best time to have an automobile accident, that human eyes work better in stronger light. No, glimpses of movement at twilight prove that Nessie is rising to the surface to forage. It's a shame to be skeptical of such a demure sea monster, one who clearly has a personality as shy and crepuscular as William Brownell's. But it seems likely that the border habitat where Nessie cavorts is not twilight but rather the equally misty realm where the imagination faces nature and wonders what the hell it just glimpsed.

It had grown darker as they talked, and the wind was sawing and the sawdust was whirling outside the paler windows. The underlying churchyard was already settling into deep dim shade, and the shade was creeping up to the housetops among which they sat. "As if," said Eugene, "as if the churchyard ghosts were rising."
—Charles Dickens, *Our Mutual Friend*

The Twilight Zone

A great admirer of the mysterious and elusive, Rod Serling was already the successful author of such Emmy-winning teleplays as *Requiem for a Heavyweight* when he was finally able to launch his own anthology series on television. *The Twilight Zone* premiered in 1959. Serling went on to win more Emmys than anyone else in history and to develop a cult following that lasted long after the show was canceled. Fans of the original series would run for the television set as soon as they heard the first few notes of Marius Constant's brilliant theme music, which people still imitate when trying to suggest eeriness.

The pilot was the third that Serling had written but the first that was tame enough for CBS's marketing wonks to approve. Television was then the most timid of America's entertainment media. Eisenhower was president; TV was black-and-white and so were the views of its censors. But Serling wanted to explore the infant medium and address some of the serious issues that TV usually ignored—aging, nuclear war, racism. He achieved his goal in the time-honored fashion of storytellers everywhere: by exaggerating his points rather than watering them down, by turning beauty into ugliness, sending the elderly to extermination camps, portraying the unreasonable fear of the Other via encounters with aliens. Censors, predictably literal-minded, considered his stories harmless fantasy.

Because of long-established associations with this time of

day, Serling's program is relevant to our theme as we explore the borderland between daylight and darkness. He could not possibly have called his series *The Dawn Zone*, even though dawn and twilight share similar light levels. In our minds dawn is about possibility and new beginnings, about the approaching light of reason. Nor would Serling have tried to convey fear or creepiness with the title *The Noon Zone*; broad daylight vanquishes the most resilient spook. Even *The Midnight Zone* would not have worked. The unequivocal witching hour is not the place for uncertainty or moral ambiguity, for uncanny situations that may or may not turn out to be supernatural.

Twilight, in contrast, is woven of uncertainty. At this time of day we feel the impending departure of light and comfort and warmth, hear the restless stir of dangers lurking in the growing darkness. We walk or drive more quickly in order to reach the safety of home before darkness falls. Just as morning symbolizes spring and youth, so has twilight come to represent late fall and old age, the foyer to winter and death. In the arts it seems that we have clear ideas about what to expect from daylight and nighttime, but we worry that at twilight anything can happen.

There is a long twilight in these hills. The sun departs, but the day remains. A sort of weird, dim, elfin day, that dawns at sunset, and envelops and possesses the world. The land is full of light, but it is the light of no heavenly sun. It is a light equal everywhere, as though the earth strove to illumine itself, and succeeded with that labor. . . . It is a world that we do not understand, for we are creatures of the sun, and we are fearful lest we come upon things at work here, of which we have no experience, and that may be able to justify themselves against our reason. And so a man falls into silence when he travels in this twilight, and he looks and listens with his senses on guard.

—Melville Davisson Post, "A Twilight Adventure"

Three Kinds of Twilight

Because human beings cannot *not* analyze and categorize, we have actually subdivided the fleeting natural phenomenon of twilight into three species—civil, nautical, and astronomical. Their official definitions are rather too precise for most of us to make use of, but a glance at them helps us see this elusive time of day more clearly. The United States Naval Observatory defines *civil twilight* as a period that occurs in both morning and evening, when the center of the sun is six degrees below the horizon. At the end of civil twilight in the morning, and at its inception in the evening, the horizon is well defined and, under normal atmospheric conditions, the higher-magnitude stars are visible. We don't necessarily require artificial illumination at this time, although we did shortly before (this morning) and we soon will (this evening).

Obviously the reverse of the evening's twilight occurred earlier today at dawn, so let's concentrate on the one we are now experiencing. *Nautical twilight,* according to the rulebooks, begins when the center of the sun has dropped to twelve degrees below the horizon. Human eyes may still be able to distinguish larger objects, but smaller and darker ones have begun to fade into darkness. The horizon is indistinct, and in most activities artificial illumination is called for. The last stage, *astronomical twilight,* ends when the sun is eighteen degrees below the horizon, after which point it contributes no more light to the sky. This point, the lovely moment when the sky is often—very briefly—an impossibly rich, almost but not quite black indigo hue, is the beginning of full darkness. This is the moment when twilight ends and true nighttime begins.

Memory Has Left the Sky

Memory has left the sky. It is night.

—Yvonne Vera, *Under the Tongue*

Afraid of the Dark

As indecisive twilight finally surrenders to darkness, the birds that clamored for attention at dawn are winging silently homeward. It is now dark enough that you can no longer be certain that the alarming scurry in the air just above your head is a bat zeroing in on twilight insects. Perhaps it's a demon. As soon as the sun goes down we recognize a new emotion in our saunter through the cycle of the day: anxiety. What will darkness bring? What is that movement over there near the corner of the house? Isn't this a high-crime neighborhood? In darkness shrubs are doubled by their shadows, cars reduced to featureless shapes behind the blinding glare of headlights, colors turned alien.

Here is the other side of Apollo's rational daylight. Evolving as diurnal primates, our clan developed imagination long ago and quickly began to think one step beyond the reality to associate darkness with evil and nighttime with death.

Mephistophilis in Marlowe's *Doctor Faustus* calls upon Lucifer, "chief lord and regent of perpetual night," to witness his blood pact with Faustus. What could better express our fear of death than the horrific idea of "perpetual night"? Sir Thomas Browne said it even better, as he said so many things so well, in his essay "On Dreams": "Half our days we pass in shadow of the earth; and the brother of death exacteth a third part of our lives."

Think about that phrase, "shadow of the earth": think of our planet's shadow, umbra and penumbra, falling as a cone into space before us as we gaze into the dark sky above. We stand on the ball that forms the base of this cone of shadow. In Shelley's long poem *Prometheus Unbound*, Earth says to the moon, "I spin beneath my pyramid of night / Which points into the heavens." From our point of view now, Earth has turned its back to the sun. In his book *Enchanted with the Night*, the Canadian poet Christopher Dewdney calculated the actual size of nighttime, in both width of the path of darkness and total square miles of darkness. Earth has a mean circumference of 24,881 miles, half of which is 12,440 miles. Subtracting those borderlands of night that are not fully within nautical twilight, the full darkness that starts about half an hour after sunset, you find that the nighttime enfolding you is reaching 11,400 miles around the world. With a surface area of roughly 197 million square miles, precisely half of Earth, or 98,500,000 square miles, is dark at any moment. Again subtracting the twilight edges, you discover that 90,292,000 square miles of our globe are in total darkness at any one time: billions of people, countless animals, under the stars. Supposedly the world turns at the same speed in the dark as in the light, a constant of physics that merely proves the relativity of passing time, because long, lonely nights are a staple of pop songs, and when you think of the size of night it's easy to understand why.

Briefly this morning, just before dawn, we tried to imagine the preindustrial darkness that has been the lot of human beings and other animals for most of our history on this planet. But how often in the twenty-first century, especially in the technology-wrapped urban world, does anyone experience even relative darkness? From the overlit, push-button perspective of this millennium, it is difficult to even envision it. Now, as the last of twilight fades into night, the reason why is all around you: everywhere you turn, from shop windows to automobiles and streets, lights are blinking on like weapons drawn to protect us from the enemy darkness.

"Then comes that mysterious time," wrote Nikolai Gogol, "when lamps endow everything with some enticing, wondrous light." Gogol was adept at seeing what everyone else overlooked, but in this case he really was describing a new experience that he was sharing with his fellow Russians. He was writing in the early 1830s about Nevsky Prospect, the historic boulevard in St. Petersburg. No aspect of the industrial revolution in the nineteenth century changed our daily experience of life more than the widespread distribution of public lighting. The use of gas street lamps caught on quickly and began to change the tenor of nighttime, ahead of many advances in medicine, communication, and transportation, that trinity so often thought of as identifying the modern age.

As early as the 1500s, however, many homes in European cities had been required to hang a lantern out front to identify the house and help illuminate the public thoroughfare. Not until the late seventeenth century did city authorities begin to place lamps along the street and maintain them for the public good. The Sun King, Louis XIV, ordered that Paris streets be hung with lanterns suspended from cables and centered over the boulevards, a forerunner of both streetlights and traffic lights. Under the strict codes of early modern urbanization, the forerunners of police were responsible for this part of the infrastructure, because light was considered the first and most essential level of security. It still is. All too often, an electrical outage—whether from hurricane or power plant failure or bomb—results in looting and violence under cover of darkness.

※

Long ago, our ancestors decided that the nightly departure of light—along with our clawless hands and weak muscles and slow locomotion—was a natural barrier that they were unwilling to meekly accept. One of the first concepts that distinguished the clan that would later call itself *Homo* was the brilliant idea that illumination and warmth could be stolen from a lightning-struck tree. If the simplest and most economical response to the daily cycle back then had not been to retire at nightfall and rise at dawn, we would not have the image of an industrious student burning the midnight oil. Most of our history was written by candlelight or whale oil, long before the Industrial Revolution mass-produced lamps and finally electrical lights. For countless millennia, when the sun went down and the shadows gathered, our ancestors fought back with whatever light they could find, just as they bundled against the cold.

Daily we repeat this ancient experience. On your drive home from work this evening, you experience the odd moment when, as your jaded gaze skims across a row of dark houses on a residential street, you see the first light come on in a kitchen window. You find yourself momentarily drawn out of the stress of attentive driving, out of the workday's accumulated weariness, as you envision the domestic life of other human beings and suddenly remember—it's a shock sometimes—that they are just as real as you are, carrying just as much of a day's experience into the night.

Artificial lighting has accompanied your entire day, of course, from shower stall to office elevator, but ubiquity breeds contempt. As you walked toward your car a little while ago, watching the setting sun romanticize the cityscape, streetlights came on around you, but you didn't notice. As you drove down the highway at dusk, you realized that it was time to turn on your car's headlights, and you saw their reflection materialize in the rear bumper of the car ahead of you, but the familiar scene didn't wake your attention. Yet this single light in an otherwise dark house, casting its bright rectangle of light as if your gaze activated it, now makes you aware that darkness is being automatically kept at bay all around you. We park under a light at home and go indoors to an evening of artificial lighting that doesn't cease until we're actually lying in bed, and even then there's likely to be a nightlight in the hallway.

For at least a few thousand years, we have had some form of lighting technology beyond campfires—fireplaces, rush-lights, torches, candles, lanterns, oil lamps. Even if we experience a power outage and find ourselves groping through dark rooms full of mischievous furniture, we don't worry because in a moment we will be rescued by a portable battery-powered light source. Scientists insist that this banishing of natural darkness is a primary contributor to a widespread health problem: insomnia. Home designers and health advocates advise us to hang dark curtains in the bedroom to keep out the dawn, so that we can balance out the hours of natural rest that we lose in the artificial lighting that tries so hard to prolong the day.

Thunderbolt

Dolphins are afraid to leap into the air but yearn to escape the boiling water. Dead sea-calves float on the roiling seas. Three times Neptune tries to emerge from the ocean, but the heat of Phaethon's veering sun chariot is too much for him, and he ducks his trident and streaming beard back under the water. The seas themselves rush to hide in underwater caverns. Finally Mother Earth raises her scorched face, with an arm held up to shield her eyes. She shudders and all the globe quakes—palaces, villages, mountaintops.

"Great Zeus!" she cries. "Lord of us all! If I deserve this burning fate, please don't spare me your thunderbolts. Slay me now with lightning instead of fire. Is this the patient earth's reward for fertility, for crops and leaves and the incense that perfumes your own altars? And even if I somehow deserve your wrath, why must the waters suffer? Surely your brother Neptune has not somehow offended you. Your own airy kingdom is not safe. Smoke fills the air and flames may topple the vaults of heaven. Even Atlas suffers. The red-hot axis of the earth already scorches his back. Zeus, my lord, if earth and sea and air die in flames, all the universe has left is primordial Chaos. Did we fight the Titans for an Order that we then abandon to the irresponsibility of a child? Save the cosmos from this fate!"

Earth withdraws to hide in her deepest caverns, as Zeus convenes a senate of the gods—especially Apollo, whose family indulgence has endangered Olympus itself. "If I do nothing now," thunders Zeus, "the earth burns and the heavens burn with it." He climbs to the highest peak of heaven, above the clouds, up to where he himself dispatches clouds to shield the earth below. This is the realm of his servants, thunder and lightning. He hefts a thunderbolt, balances it, draws back his sinewy arm, and flings a bolt across the sky— sending fire to end the fire that is ravaging the earth.

Appropriately for the son of fire and water, Phaethon dies in a brilliant flash of light and falls toward the sea. The few brave peasants on the seashore hear a thunderclap and look up in time to see Phaethon's corpse tumble through the air, trailing smoke. The chariot begins to crumble, its axle and wheels and broken fragments scattering through the sky and then falling toward earth. Phaethon's body falls into the great river Po that feeds the Adriatic, far from his home. The Naiads, sea nymphs, bury the boy in view of the water. They carve a memorial on his gravestone: "Here Lies Phaethon, child of the Sun. He dared to command his father's chariot. He failed but died bravely."

Apollo knows that his son was reckless more than brave, and that the father is more to blame than anyone. Lands that were gardens are now deserts; rivers have run to hide in faraway mountains. Some creatures now lurk in caves or in the depths of the sea and may never again return to sunlit Earth. Apollo knows that Mother Earth spoke the truth: that Order vanished and Chaos almost returned because the sun strayed from its course. In despair for his loss and guilt, Apollo buries his face, and for the first time since the gods crafted the universe, for one entire day the sun does not rise and the world remains in darkness.

The Changing of the Guard

Most of our attention today has been upon external phenomena and what we have thought about it. Now that darkness has replaced daylight, it is time to turn our attention inward, to glance at a couple of the many ways that we ourselves are influenced by the rhythms of the day and night. About the time that the sun goes down, the human body's systems of

nerves, hormones, and other interconnected support machinery begin the changing of the guard. The process occurs
slowly, in response to gradual changes in the light. Our innate
bodily response to the cycles of light and darkness date so
much further back in time than clocks and calendars that
they make all of human history seem to occupy only a
moment. We are psychologically accurate if we say that these
actions have been recurring since the beginning of time,
because the dependable passing and return of the day is our
first measurable experience of this elusive concept.

As Earth's rotation on its axis and its revolution around
the sun have fashioned the cadence of our lives, nature has
had plenty of spare time to tinker in the shop and gradually
condition its inhabitants to this ancient cycle. For billions of
years before life evolved on Earth, the planet was already tire-
lessly rerunning the alternation of daytime and nighttime,
light and darkness, warmth and cool, and it instilled this pattern in the earliest life forms. Indeed, numerous scenarios
about the evolution of the first organisms, in the warm seas
of the primordial planet, argue that another related daily
rhythm—the rise and fall of tides—may have helped nour-
ish early life.

Many kinds of bacteria have been observed to function on
daily cycles; some even possess DNA repair mechanisms activated by light. Long before there were human beings to talk
about perceptual changes at this time of day, animals were
waking and sleeping, hunting and fleeing on a natural schedule. The timetable even includes shift changes, the night crew

arriving as the day shift heads home to the burrow. Such daily cycles not only persisted over the eons but became increasingly important as animals evolved more elaborate behavior and more complex societies. Millennia before scientists measured the physiological toll on third-shift workers, even before someone daubed pigment on a cave wall to express her fear of the night, our protohuman forebears lived their lives on the ancient daily rhythms inherited from earlier primates.

The human race has always been aware of these rhythms, of course, but until the mid-twentieth century we didn't have a technical word for them. Then scientists coined the term *circadian,* from the Latin for "about" or "around" plus "day." Circadian rhythms influence everything from blood pressure to test-taking ability, from temperature regulation to insomnia. The body's response to various medications, even some essential cancer drugs, depends upon which time of day they are administered. Many late-shift accidents, including those at major industrial sites such as Three Mile Island, have occurred when work schedules fell out of alignment with our natural rhythms, demanding attention instead of sleep at three in the morning. Seasonal affective disorder, with its cutesy abbreviation SAD, is exacerbated by reduced light levels in wintertime.

Scientists have long known that pigments—the protein compounds called opsins mentioned at dusk—occur in the retina, where they absorb photons and help translate them into electrochemical signals that pass along the optic nerve to the vision-related processing centers of the brain. At this point the light that entered your cornea can no longer be called light. It is still what it was to begin with—a form of energy—but the biochemical machinery inside the eye transforms it into electrical impulses. Opsins were first discovered in the 1870s. Rhodopsin, the light-sensitive pigment in the rod cells of the eye, kicked in a little while ago when the low wattage of twilight illumination began to challenge your vision.

For many years, researchers assumed that the same kind of pigment was involved in both vision and the regulation of circadian schedules. In 1998, however, scientists announced the discovery of a previously unknown light-sensitive pigment, found in many areas of the body, that helps regulate our biochemical clock. Named *cryptochrome,* and existing in two forms abbreviated as CRY 1 and CRY 2, the pigment shows up in various tissues, including parts of the brain and areas of the skin, as well as in the eye. Researchers found cryptochrome in a different area of the retina than opsins, and it is linked to vitamin B_2 production—unlike opsins, which are involved with vitamin A. This B_2 connection has suggested to scientists that victims of SAD may carry a defective gene or that their cryptochrome is otherwise unable to manufacture the normal amount of the vitamin. In absorbing blue light and transmitting the resulting signal to a different area of the brain than the vision centers activated by opsins, cryptochrome enables the body to synchronize its circadian clock.

A clue to the distinctive functions of opsins and cryptochrome is that some blind people still maintain circadian rhythms, because their retinas still contain cryptochrome. Yet the eye-brain connection in both situations is emphasized by another curious fact: when researchers severed the optic nerve of test animals, the animals lost not only sight but also their built-in circadian regulator. Scientists have proven that cryptochrome in laboratory mice responds to light signals outside the body in order to regulate processes inside the body. It is biochemically similar to the bacteriological DNA repair kit mentioned above, as well as to the blue-light-sensitive photoreceptors that influence the growth of plants.

Lag

One important daily alteration inside the human body is behind a familiar side effect of the space age. As soon as transcontinental flight became routine, its attendant violation of daily patterns began to afflict travelers. Jet lag results from our recently acquired, and quite unnatural, ability to leap across thousands of miles of Earth in a few hours. It is still a relatively new problem, this disorientation after our magic carpet ride, analogous to illnesses such as rickets that developed aboard the first ships whose crews sailed long enough to suffer vitamin deficiencies. Long-range travel wreaks havoc with the body, as it attempts to function normally although its solar cues have been abruptly thrown out of sync. Of course, similar changes occur in the body in response to night-shift jobs or shift rotations that demand constant alterations in the sleep cycle.

Each regular international flier has a different method for

dealing with jet lag. The wealthy advise that we recline com-
fortably in a first-class seat; some insist that an in-flight
massage works wonders. Frequent fliers in coach class rec-
ommend remedies as diverse as vitamin supplements and
artificial lights—which, despite their supposedly therapeutic
effects on the person employing them, are guaranteed to irri-
tate fellow travelers. One of the simplest and most natural
ways to reduce the effects of jet lag is to remember what
causes it—a sudden violation of our daily rhythms—and
make the change less traumatic by gradually synchronizing
your bedtime with that of your destination for several days
prior to a long flight.

Nowadays many of us jet lag sufferers, complaining that
loss of sleep is our main problem on long flights, respond as
we do to other troubles—by swallowing a pill. To the long-
range flier as to everyone else, prescription sleeping pills
offer unconsciousness on demand, and in this case they can
distract you from the knowledge that you are uncomfortably
crammed into a pressurized tin can rocketing through the
upper story of the atmosphere.

But there are non-narcotic alternatives to prescription
sleep aids that cleverly respond to, rather than resist, the
body's internal cues directing its circadian rhythms. One
involves taking a small dose of melatonin, which is a hormone produced by the pineal gland. Although melatonin is
not the miracle hormone that its hype in the 1990s claimed,
it has become the most popular treatment for jet lag and
seems to assist in treating seasonal affective disorder. Studies
confirm that small doses of melatonin taken earlier each

night for a few nights prior to a flight will help the body adjust the sleep cycle toward its new destination. Scientists are still analyzing the hormone's several roles in maintaining our daily rhythms, but the news media have been quick to promote a simplified equation: melatonin makes us sleep and serotonin makes us wake up.

The body produces melatonin during the night—more, the longer the night, biochemically adjusting your body's systems to changes in length of day throughout the year. Many other species' bodies work in the same way. The pineal gland responds to directions from the suprachiasmatic nuclei of the hypothalamus, the versatile hormone factory in the brain that also produces hormones controlling such functions as appetite and body temperature, as well as those that influence emotions and inspire the urge to reproduce. This pattern of hormone production forms the basic internal clock and calendar for many species' seasonal rhythms. It is one of the natural functions with which artificial light can interfere. A century ago most people not involved in war or burglary slept nine hours per night. A 2002 study reported that twenty-first-century urbanites and suburbanites, even with no livestock to feed and a houseful of labor-saving devices, slept more like seven hours, with a high percentage reporting only five or six hours.

Melatonin production is inhibited by exposure to light. As the sun comes up, the pineal gland produces less melatonin and the body begins producing serotonin. (This neurotransmitter, which plays a role in inhibiting sleep and a number of other bodily functions, is produced by the brain and spinal cord, but also acts as a hormone when produced by cells in the lining of the digestive tract.) Melatonin has the ability to send feedback to the pineal gland, thus participating in modifying its own production levels and responding to changing light at different times of day or year. Daytime lethargy and unaccountable sleepiness—which can be not only depressing but quite dangerous in many jobs—often turn out to be influenced by a disproportionate or otherwise skewed production of melatonin, in too large amounts or at incorrect times of day.

The recent surge of interest in circadian rhythms in the human body—especially in sleep disorders—dates back to two studies conducted in the early 1980s, one on human beings and one on rats, which indicated that melatonin could shift the phase of our internal clock. Sometimes carefully regulated and timed lighting can nudge melatonin production upward or downward. In more ways than we can imagine, the life and culture of *Homo sapiens* is influenced by the electrochemical lightning in the brain as it responds to light and dark, and to subtle responses that occur throughout the body.

Some research indicates that, in pregnant women, melatonin production, which clearly varies in proportion to daylight (and therefore by season), may subtly influence the development of the fetus's brain pathways. Researchers who interviewed preschoolers in both the Southern and Northern hemispheres—in New Zealand and the United States—noted a dramatic correlation between the children's birth date and their relative shyness. During mid-gestation the neurons in the cerebral cortex are shaping into the crucial patterns of the brain. In the Northern Hemisphere those children born in autumn and early winter, and whose mothers therefore experienced mid-gestation during late spring and early summer—periods of more light—were rated as considerably less inhibited than those born in spring. The opposite pattern prevailed in the Southern Hemisphere. Of course, plenty of other factors may have been involved, but now and then scientists stumble upon the mysterious rationales for what have long been considered merely superstitious connections between birth date and personality.

In 1989 a young Italian woman named Stefania Follini spent 130 days—more than four months—in a sealed cave in Carlsbad, New Mexico. She had volunteered for an experiment focused on learning how the human body might respond to long periods of isolation during interplanetary travel. While underground, she slept ten hours at a time but often stayed awake for up to twenty-four, and her menstrual cycle completely stopped. She had so lost track of days that when Maurizio Montalbini, the Italian sociologist who directed the experiment, contacted her via computer to alert her that her four months were up, she was shocked. She thought she had been underground for only half that time.

A few years later, Montalbini himself performed an even more dramatic experiment showing how our disengagement from solar rhythms influences our sleep cycle and our perception of passing time. He spent an entire year in a cave. In the world above, 366 days and nights came and went while Montalbini stayed underground. One of his primary goals was to record his subjective response to the loss of solar cues, which meant that, like Follini, he took no clocks or calendars with him. When fellow scientists came for him after his year completely removed from the natural cycle of sunlight and darkness, he thought he had spent only 219 days underground. But in October 2006 he entered another cave, this time planning to spend three years away from the regulating light of the sun.

Time makes everything old so the kissing, young darkness became a monstropolous old thing while Janie talked.

—Zora Neale Hurston, *Their Eyes Were Watching God*

Darwin's Busy Plants

Some of the early clues about circadian rhythms were first noticed in other organisms rather than in human beings. We can see them all around us. With the sun gone for the day, some of its silent attendants have closed up shop for the night. Keyed to the alternating rhythms of daylight and darkness, responding largely to changes in the intensity and angle of the light, many plants fold their leaves or blossoms at dusk or after dark. Sun-following plants abound. You can observe this kind of movement in your office philodendron, in roadside chicory, as well as in many blossoms in your garden. The sunflower, which earned its name by turning its head to follow our local star's arc across the sky, was once an emblem of the Inca sun god. Not surprisingly, over the centuries we have personified not just our fellow creatures but plants too, especially those that move. We have demoted this particular natural relationship from cosmic to kitschy, with such recent cuteness as images of sunflowers wearing sunglasses. We don't personify any less than our ancestors; we just do so in commerce and in parody more than in reverence.

Even though they can't deliberately change their location, plants—and, in some cases, sedentary animals, such as the adult stage of barnacles—can move at least some part of themselves toward or away from stimuli. This movement, usually in the form of directional growth, is called *tropism*. When plants' root systems inch their way into broken water pipes, they are demonstrating hydrotropism. There is even a word for a plant's natural tendency to bend away from the earth: *apogeotropism*. There is another motion in plants called "nastic movement," a response that is independent of the direction of the stimulus—for example, *seismotropism*, a reaction to shock. *Heliotropism*, also called "solar tracking," is the movement many plants exhibit in response to sunlight. This sort of automatic action lent its name to the brilliant color of one of the more restless sun-loving plants, the flower called heliotrope; and, because language itself mutates like living things, we now have crayons and bath towels whose color is described as heliotrope. Scientists distinguish between turns perpendicular to the sun's direct rays (*diaheliotropism*) or parallel to them (*paraheliotropism*).

With plants swaying all around us in the declining light, revealing that they too join in the dance of sun and Earth, it is a good time to talk about the pioneer studies in circadian rhythms performed by that versatile and surprising man Charles Darwin, who wondered if plants sleep.

✻

In our era, despite the whining of creationists—and their flavor of the week, intelligent designers—Charles Darwin is achieving something of the pop-star status that made Albert Einstein such a trendy figure in novels, plays, and even movies of the 1990s. Biographies and studies proliferate. Already Darwin and his once revolutionary ideas have starred in such excellent literary works as *Mr. Darwin's Shooter*, by the Australian novelist Roger McDonald. For many years, Darwin's bearded face has even adorned the British £10 note.

Every literate human being is aware of the concept of natural selection, the process by which animals and plants gradually adapt to their environment. It is the idea behind most of modern biology, as well as the essential tenet of conservation: the interdependence of creature and environment. These notions were first expressed at length in Darwin's 1859 magnum opus, *On the Origin of Species by Means of Natural Selection, or the Preservation of Favoured Races in the Struggle for Life.* Eleven years later he published the first book to deal directly with our body's evolutionary kinship to other animals, *The Descent of Man, and Selection in Relation to Sex.* He is also remembered for one of the great travel volumes, his 1839 *Journal of Researches into the Geology and Natural History of the Various Countries visited by H. M. S. "Beagle."*

Such wide-ranging volumes, however, don't represent Darwin's total contribution to and influence upon biology. Most of his days at Down House—his home in Kent, formerly south of London and now in its suburbs—he spent working on the seemingly smaller projects that provided much of the data for his larger thinking. For eight years he dissected barnacles before writing a comprehensive tome about them, and his final book, published the year before he died in 1882, was *The Formation of Vegetable Mould, through the Action of Worms, with Observations on Their Habits.* He wrote several books about plants, their topics including the ways that insects fertilize orchids, the effects of cross- and self-fertilization, and the various forms of plants of the same species. Not as scandalous as his other work, these books were largely ignored by the general public, even as the old man's reputation grew to the point that he was buried in Westminster Abbey (where, early in the next millennium, an American creationist was caught spitting on his grave).

In 1880, half a century after his departure on the round-the-world voyage of the *Beagle,* Darwin published the book that interests us most in the context of our journey through the day, *The Power of Movement in Plants.* It is a quantitative, analytical tome, the result of decades of observation in the field and in his home laboratory and greenhouse, as well as the culmination of years of correspondence with scientists and farmers in many countries. At six hundred pages, including almost two hundred woodcuts, it is his largest book about plants. Even in a multiple-choice list, most people would be unable to link this book to Darwin, yet while introducing a

new edition in 1966, the American botanist Barbara Gillespie
Pickard maintained, "If asked to choose the scientific report
which most clearly marks the beginning of the modern study
of plant growth, a great many botanists would select" this piv-
otal volume. In our own time, research on the topic is more
likely to appear in a peer-reviewed journal under a title such
as this one from *Plant, Cell and Environment* in 2005: "Circadian regulation of leaf hydraulic conductance in sunflower
(*Helianthus annuus* L. cv Margot)."

Darwin was in his late sixties as he began his three years
of work on this book. Plagued by lifelong health troubles, he
was growing ever more frail. Yet even while lying ill and
groaning on a sofa, he didn't spend his time reading *Punch*
or the latest triple-decker novel by Bulwer-Lytton—who, incidentally, once parodied the great scientist as the eccentric
Professor Long, a man obsessed with the private lives of
limpets. Instead Darwin watched how plants' leaves turn
toward the sun and how they behave in darkness.

Working with his grown son Francis, a botanist and later
his father's editor and biographer, Darwin performed and
recorded thousands of experiments involving hundreds of
plants. He concluded, among other insights, that plants move
via changes in osmotic pressure within their vascular system,
and that "sleep" movements protect them against too dramatic radiation of heat outward from their leaves during the
night. He monitored the movements of vines as they fumble
for a supporting object to wrap themselves around and the
way that roots grope past stones and other obstacles in the

soil. He studied how some plants spread their leaves toward the rising sun. He wondered about the mechanisms by which plants respond to the slightest external pressure. How does a light touch on its toothed mouth trigger a Venus flytrap to snap shut to capture an insect and send it tumbling into the plant's sticky throat to be digested, like an animal eating prey?

Darwin was fascinated. His gee-whiz appreciation of the natural world, back when he was dissecting barnacles, had inspired his affectionate children to say that he sounded like a salesman. Typical of this attitude was his remark to the American biologist Asa Gray back in 1863. He observed of the sensitive tendrils of certain plants that "their irritability is beautiful, as beautiful in all its modifications as anything in Orchids." He goes on to describe his point in more detail. About the spontaneous movement of certain plant tendrils, Darwin asked Gray: "Pray tell me whether anything has been published on this subject? I hate publishing what is old; but I shall hardly regret my work if it is old, as it has much amused me."

Many of Darwin's experiments were masterpieces of simple and elegant thinking, although to us they seem quaint. He and his son graphed some plant movements by the simple expedient of tracing the outline of a leaf periodically during many hours of sunlight and darkness. For example, Darwin placed one sheet of glass in front of and another above a tobacco plant whose slow daily gyrations he wished to record. As a fixed guide for his homemade graph, he drew a black dot on a white card and placed the card on a support behind the plant, securely fastened within the pot so that its position relative to the pot and the plant and viewer would not change. Then he placed a tiny, barely visible, and almost weightless dot of ordinary sealing wax at the end of a minuscule glass filament that he had cemented to the tip of a leaf about halfway up the plant stalk. "The weight of the filament," he assured his readers, "was so slight that even small leaves were not perceptibly pressed down."

Father and son took turns marking on the glass plates, with that sophisticated instrument the "sharply-pointed stick dipped in thick India ink," the ever-changing location of the wax bead. They then traced the constellation of black dots onto paper and connected the dots, with arrows, to graph direction and rate of growth. Darwin also cleverly used smoked glass as a drawing medium in which growing root tips could sketch a self-portrait of their own restlessness. He carefully took these measurements in all kinds of light, at every time of day. He kept some specimens in absolute darkness except during his moments of measuring. He reported on 141 plant varieties that exhibited some form of nighttime change in behavior that he categorized as "sleep." Many species moved rather wildly, in twists and long graceful waves or even in sudden clocklike jerking motions.

In contrast to more elaborate gyrations he had recorded, Darwin found himself almost aesthetically pleased by the simple movements of a single leaf on a tobacco plant during a sixty-five-hour period. "The tracing is remarkable . . . from its simplicity and the straightness of its lines," he wrote. His careful diagrams, showing precisely where the leaf was at, say, 3 p.m. one day and 8:10 a.m. the next, portray a simple large ellipse. The plant reached its highest point late morning, between ten and eleven, and sank to its lowest between three and five in the afternoon.

In 1881 Darwin wrote in his autobiography, which was not published until after his death, about his larger goal in these plant studies. He insisted that "in accordance with the principle of evolution it was impossible to account for climbing plants having been developed in so many widely different groups unless all kinds of plants possess some slight power of movement of an analogous kind." And he calmly added, "This I proved to be the case." Once again, while looking for adaptive measures, the great biologist was simultaneously exploring other territory as well. His quaint and primitive experiments with plants are now considered some of the most important early work on circadian rhythms, and constitute a pioneering glimpse of how the plants around us respond to the cycles of day and night.

In our time this sort of work would require grant money, laboratories, and grad-student slave labor. Darwin was independently wealthy, but his was not yet the era of big-budget science. It's entertaining to think about how much the Victorian naturalists accomplished with so little specialized equipment. Darwin's work with plant movement reminds us that the discovering is always done by the brain behind the equipment, just as not even the most elaborate camera can take a truly artistic photograph by itself. And think of all those waiting stars up there, and of the shelves of discount stores stocked with more advanced telescopes than the very best that were ever available to Galileo.

You can think better in the dark, if you've got anything to think about.

—Mati Unt, *Things in the Night*

Pillars of Light

There is no better time for seeing the world anew than darkness, and no better way to do so than on your own two legs. Sometimes on cold winter nights an urban walker can witness an entertaining atmospheric phenomenon akin to haloes, sun dogs, and the green flash. As we saw at sunset, diamond dust—particles of ice crystals suspended in the air—can cause the appearance of a halo around the sun or moon. On chilly nights these microscopic specks of ice can produce a different kind of eerie sight. They float invisibly in the air unless light hits them at just the right angle. As you walk toward a streetlight on an otherwise dark road, you may find that suddenly the electric lamp seems to cast a pillar of light upward into the night sky. The formerly invisible ice crystals have all turned into mirrors.

The attentive walker can see a related quirk of light on foggy nights, a phenomenon as surprising as the twilight rainbow. When the right number of droplets of water are suspended in the air, a streetlight sometimes casts the cruciform shadow of a telephone pole onto the heavy air itself, where it seems to float in ghostly suspension rather than obediently sprawling on the ground where it belongs. It may have taken noontime shadows to enlighten Eratosthenes and E. E. Barnard, but a shadow falling on the air itself is proof of the magical nature of night.

Pale Fire

> *The line of the horizon was clear and hard against the sky, and in one particular quarter it showed black against a silvery climbing phosphorescence that grew and grew. At last, over the rim of the waiting earth the moon lifted with slow majesty till it swung clear of the horizon and rode off, free of moorings; and once more they began to see surfaces—meadows wide-spread, and quiet gardens, and the river itself from bank to bank, all softly disclosed, all washed clean of mystery and terror, all radiant again as by day, but with a difference that was tremendous. Their old haunts greeted them again in other raiment, as if they had slipped away and put on this pure new apparel and come quietly back, smiling as they shyly waited to see if they would be recognized again under it.*

> —Kenneth Grahame, *The Wind in the Willows*

The Parish Lantern

There it is, the moon, rising above the cityscape—momentarily, and to its own embarrassment, mistaken for a streetlight on those nights when it rises white and round. Whether full or partial, orange or chalky, surprisingly large near the horizon or shrinking as it climbs, if the moon is onstage it plays the lead in the show. Just as the sun reigns over the blue sky of daytime, so does its regent dominate the night. If tonight's were its first sighting, we would all be jostling in the streets and staring upward, gesturing. Children would cry out, "Look! A rock in the sky!" The superstitious would pray, the scientists conjecture, and entrepreneurs would be selling robes for the Apocalypse. But the moon is there reliably, every week, waxing and waning, so we no longer notice it exiled above the cop shows and the auto commercials.

Long before we tamed fire, and long afterward, the moon's illumination was the only way to move around safely by night. This dependence is one of the many aspects of the past that we find almost impossible to recapture—a sense of the crucial, personal relationship one felt with such natural rhythms. The phases of the moon appear on calendars because they were once so important to everyone, and they still matter to farmers and many others. For thousands of years, even brief local outings were timed to coincide with a full moon and might be postponed at the last moment if clouds hid it from view. Sailors prayed for cloudless nights.

Because of its location in relation to Earth, in its other phases the moon is visible for only part of the night, but when full it rises at nightfall and doesn't set until dawn. When the moon rose to illuminate a summer field, tasks such as planting crops and mowing hay could be continued past nightfall. Near the autumnal equinox, the small angle of the moon's orbit makes it available in the sky longer than usual, enabling farmers to work later and leave fewer ripe crops outdoors to tempt animals and thieves. Earlier cultures named their months by the nighttime activities that the bright moon made possible—harvest moon, hunter's moon.

Because darkness welcomed the parade of brigands that has always infested civilization, a moonlit night meant fewer guards had to be posted around flocks or silos. Moonlight on snow could rival the day's illumination and permit wood-choppers, thatchers, and others to continue working. Wary thieves stayed home in bed when "the parish lantern" was at its brightest. As late as the seventeenth century, Samuel Pepys preferred to venture forth at night into London only when guarded by "brave mooneshine."

Naturally our vital satellite was quickly promoted to a symbolic level. Orthodox Muslims begin Ramadan, the ninth month of the Muslim calendar and the annual commemoration of the revelation of the Qur'an to Muhammad, with a day of fasting and prayer that begins as soon as the observer sees *hilal,* the crescent moon. Many people are exempt—preadolescent children, nursing or pregnant or menstruating women, the mentally or physically ill—but all other believers must observe the fast. Originally it was to begin as soon as the observer saw the crescent moon. Nowadays Muslim radio and television stations and print sources provide the times of the fast.

The crescent moon was an emblem of the Sassanids, the last dynasty to rule Persia before the Arab conquest in the seventh century. The Sassanid official religion was Zoroastrianism, but in time their lunar symbol evolved into a special representation of the Ottoman Empire. Many peoples have employed the crescent moon as a symbol. The South Carolina flag bears a palm tree and an oddly stylized crescent moon that looks more like the sun undergoing eclipse. During the nineteenth century, the twilight years of the crumbling Ottoman Empire before its dissolution in World War I, other Muslim nations began to incorporate the crescent moon as a symbol. Today it is particularly associated with Islam, as evident on flags from Tunisia to Uzbekistan.

The moon lurks everywhere in Japanese prints—above cherry blossoms in an evening street scene, over a commuter boat on the Yodagawa, behind bamboo leaves in the folds of a geisha's kimono—but usually it seems more celebrated than invoked. A good example is the work of Yonejiro Yoshitoshi, the nineteenth-century heir to *ukiyo-e* masters Hokusai and Hiroshige. Like many other Japanese artists, he employed the moon in artistic tableaux. But Yoshitoshi went further; for a while he even called himself Tsukioka Yoshitoshi, incorporating a Japanese word for moon, *tsuki*.

Series of prints on a related theme were popular at the time, ranging from Hiroshige's *Sixty-nine Stations of the Kiso Highway* to Yoshitoshi's own *Twenty-eight Murders with Verses*. In time Yoshitoshi produced a gorgeous series entitled *One Hundred Aspects of the Moon*. Earth's familiar satellite appears in every phase and in scenes both dramatic and lyrical. It watches a thoughtful lover and a late-night scholar, but it also illuminates murder and suicide—a creeping assassin, a soldier preparing to commit seppuku. Folklore animals wrestle and race under the moon. The great haiku poet Basho appears in this series and so does the pioneer novelist Murasaki Shikibu. Sometimes the moon itself isn't visible but its light appears—reflected, for example, in the water into which a young woman dips her foot. Like the real one, Yoshitoshi's moon watches over an entire world.

A Rock in the Sky

How odd to think that this moody, soft light tonight is borrowed from the blinding sun. "Queen of mirrors," the British poet John Wain called the moon. Timon of Athens, in his famous speech about how everything steals, declares that "the moon's an arrant thief, / And her pale fire she snatches from the sun." Vladimir Nabokov's poem/novel/mock monograph *Pale Fire* derives its title from Shakespeare's play, although Nabokov slyly makes his self-serving commentator, Kinbote, unaware of the source while frequently demonstrating stolen reflections himself.

Moonlight is indeed secondhand sunlight, Artemis's weak mirror of her brother's fire. Not surprisingly, the goddess of the moon was the twin sister of Apollo, a daughter of lord Zeus and Leto, who was in turn the daughter of primordial Titans. In most stories Artemis is a virgin huntress, quick to take offense and quick to draw her bow. Selene, a personification of the moon, merged in time with Artemis, and both eventually became identified with the Roman goddess Diana. But the Greek root remains in the names of the mineral selenite and the element selenium, not to mention such narrow specialties as selenodesy, the careful measurement of the moon's surface, a profession that in a sense began half a millennium ago, the first time that Galileo trained his telescope on the moon.

The Archangel's Telescope

Earlier today, as we explored our natural and cultural response to the sun's unfailing reappearance every morning, it was easy to realize why dawn and sunrise genuinely feel like a rebirth. Hence Marcel Proust's remark, quoted earlier today, about the earth being re-created anew every day. This idea, however—this sensation of standing in the presence of the gods and witnessing their primal creativity—is certainly not limited to morning. We can seek it at different times of day for different reasons.

Opportunities to do so await us outdoors every evening. Ralph Waldo Emerson remarked to his journal in 1841 that no matter how he tired of his surroundings, the moon, "apologist of all apologists," somehow magically "hides every meanness in a silver-edged darkness." In an essay he made even grander claims for the rock in the sky: "The man who has seen the rising moon break out of the clouds at midnight has been present like an archangel at the creation of light and the world."

Discovery can be as exciting as creation, and in psychological impact it must be a similar process. No mortal was more present at the creation, no human being more archangelically appreciative as the moon broke out of the clouds, than the first man who wrote about looking at Earth's satellite through a telescope. With the sun new in the sky this morning we considered the painstaking discoveries of Copernicus, his rejection of Ptolemy's geocentric model of the cosmos and his own elucidation of a heliocentric system. The church ignored these supposedly blasphemous claims until 1616, several decades after the death of their discoverer. It responded then because of widespread attention to discoveries by another important figure in the history of skygazing, the Italian astronomer Galileo Galilei—discoveries that went a long way toward documenting the heretical claims of Copernicus.

※

Galileo was born in 1564, two months before Shakespeare, to a distinguished but impoverished Florentine family. He resisted his father's urging toward a career in medicine because dutiful recitation of Galen and Aristotle, unaccompanied by any attempt at real discovery, bored him. His challenging questions quickly gained him the reputation of a troublemaker. Eventually he turned toward other subjects, and by the age of twenty-eight he occupied the chair of mathematics at the University of Padua. His science always had a practical bent. In the late 1590s he modified the existing proportional compass used by engineers all over Europe, and soon his new version was in such demand that he hired a craftsman to produce it in quantity. When someone else claimed credit for his invention and plagiarized his manual of instruction for it, Galileo's public cannonade launched a lifetime of growly polemic, not to mention endless complaints about lack of money. In time he also became tutor to Cosimo II, heir apparent to the Medici dynasty. Like Leonardo a century before, Galileo was a versatile man. He outlined the physical laws involved in the pendulum and designed a prototype pendulum clock. He discovered sunspots and the satellites of Jupiter. He designed an irrigation device and studied the trajectory of cannonballs. But most relevant to our journey tonight, he trained a telescope on the moon and described with the astonishment of an explorer what he found there.

In 1609 the German astronomer Johannes Kepler had published *Astronomia Nova* (The New Astronomy), the last great work of naked-eye astronomy. Unfortunately for his contemporaries but to the delight of later historians, he buried his true discoveries amid long-winded accounts of false trails upon which he had wasted time, and the work was slow to gain recognition; it didn't fully prove its worth until Newton came along. Meanwhile Kepler, using the detailed studies of planetary movements recorded by his mentor Tycho Brahe, especially Tycho's voluminous notes on Mars, further undermined the notion of a perfect and unchanging celestial realm. In his now classic work *De Stella Nova* (Of the New Star), Tycho had challenged this daydream with his conclusions about novae and how they appear out of nowhere among the supposedly fixed stars. Kepler found more flaws in the ideal heavens. He incorporated Tycho's data about the wandering planets into his now legendary laws of planetary motion, arguing that discrepancies in earlier models resulted from attempts to force imaginary circles onto what in reality were elliptical orbits. In the minds of classically trained theologians, an ellipse was scandalously less perfect than a circle.

Galileo was not yet aware of Kepler's contribution, which would soon unite with his own to help overthrow the old intellectual regime. During 1609 he read about a recent Dutch invention, "a spyglass by means of which visible objects, though very distant from the eye of the observer, were distinctly seen as if nearby." (The word *telescope* was not coined until 1611.) This invention has often been attributed to a Middelburg lens grinder named Hans Lippershey, but we now know that it existed prior to his claim. Some people confirmed the report and others denied it, but soon Galileo set about building an instrument of his own upon the same principles. He began by taking a lead tube and fitting it with two glass lenses, each of which had been ground to be flat on one side. On the reverse one lens was convex and the other concave; the greater the difference between the curvature of the lenses, the higher the magnification. Galileo placed the convex lens at the far end of the tube and peered through the concave lens. Objects appeared, he noted, three times closer and nine times larger than when seen with the naked eye—"satisfactorily large and near." His next telescope magnified objects more than sixty times, and the next "nearly one thousand times larger and over thirty times closer than when regarded with our natural vision." Galileo omits, and for the sake of brevity we will have to omit, other fascinating details such as the speed with which this new invention was spreading (even as he learned of its existence it was becoming a toy on the streets of Paris), and how it instantly seemed to him not only the ultimate tool for stargazing but also a path to the wealth he had long sought.

The information above comes from his revolutionary and highly readable scientific milestone, *Sidereus Nuncius*, usually translated as "The Starry Messenger" although Galileo himself seems to have intended the title to mean "A Message from the Stars." The term *sidereal period* still describes the time required, relative to the stars, for a planet or satellite to make one rotation or revolution; and *nuncius* is related to *annunciation* and *announcement*. The rest of the title page, which combines information that today would be divided into publisher's jacket copy and author's biographical note, is so informative it is worth quoting in its entirety:

> Revealing great, unusual, and remarkable spectacles, opening these to the consideration of every man, and especially of philosophers and astronomers; as observed by GALILEO GALILEI, Gentleman of Florence, Professor of Mathematics in the University of Padua, WITH THE AID OF A SPYGLASS *lately invented by him,* In the surface of the Moon, in innumerable Fixed Stars, in Nebulae, and above all in FOUR PLANETS swiftly revolving about Jupiter at differing distances and periods, and known to no one before the Author recently perceived them and decided that they should be named The Medicean Stars.
>
> Venice 1610.

Note that Galileo states from literally the first page of the book his intention to make these marvels available to the consideration of everyone. Throughout his life he wrote not merely—not even primarily—for his fellow astronomers and physicists; he wrote for everyone who was interested in learning more about our place in the universe. At the time, scientific discourses were immediately available to colleagues abroad because they were written in Latin, equally comprehensible to educated Italians and Scots and Germans. Galileo wrote not in Latin but in Italian, simultaneously making his work accessible to more of his fellow citizens and fewer of his international colleagues. But translations began to appear almost immediately. Galileo was a vivid and lucid writer, with a talent for analogy and an entertaining inability to resist lampooning those whose ideas were opposed to his own. It is unfortunate that imaginary trips to the moon such as Lucian's *Icaromennipus*, however entertaining they may be, are considered literature by many general readers who relegate Galileo's classic to the dusty shelves of mere science.

Because his observations are so revolutionary, Galileo begins his account not with conclusions or theories but with a description of first peering through a telescope at the moon. The key to his success lies in his method of permitting the reader to accompany the progress of his discoveries, and thus to find them irresistible and convincing in the same way that he did. His description of his first glimpse of the moon single-handedly refutes Keats's complaint that science unweaves the rainbow. No myth about squabbling lunar deities, no lovers' moonlit tryst, outweighs the first description of the moon's surface in either poetry or psychological relevance.

Galileo points out that for explanatory purposes he must immediately distinguish between two parts of the moon's surface: the lighter and darker areas that make up the familiar "face" of the man in the moon. The naked eyes of human beings, he insists, have been observing these features throughout history, but no one before Galileo himself had been granted the opportunity to examine their details more closely. And immediately he flings down a gauntlet before the church:

> From observations of these spots repeated many times I have been led to the opinion and conviction that the surface of the moon is not smooth, uniform, and precisely spherical as a great number of philosophers believe it (and the other heavenly bodies) to be, but is uneven, rough, and full of cavities and prominences, being not unlike the face of the earth, relieved by chains of mountains and deep valleys.

How did he come to these conclusions? He is quick to explain. In order he describes what he observed in this initial close-up glimpse of another planetary body. Armed with knowledge of scientific perspective acquired by Renaissance artists, he reports that, a few days after a new moon, when the young crescent still has sharp horns, the border of sunlight and shadow does not fall across the crescent as a smooth oval line, which is the elegant curve one would expect from a perfect sphere. Instead, wherever a lighter area meets a darker area, the line is "uneven, rough, and very wavy." Galileo provides detailed drawings. He explains the nature of the irregular line, that luminous excrescences extend well into the darker portion at various sites, and that dark spots, quite separate from the larger dark region, are scattered across the brighter face of the moon. When observed through a telescope, the dark spots all portray a similar structure, with their darkest areas toward the sun while the side opposite the sun appears "crowned with bright contours, like shining summits."

To explain his observations to people who had never gazed through a telescope, Galileo here begins a series of analogies with easily observed phenomena on our own planet. He compares the similarity of what he was watching on the moon—the changing light along what appeared to be mountaintops—with a sunrise on Earth that leaves valleys shadowed long after peaks are "ablaze with glowing splendor on the side opposite the sun." And it was unlikely that he had misinterpreted, because this was not a static phenomenon that he had witnessed. Just as valley shadows on Earth diminish as the sun rises higher, so do the dark spots on the moon shrink as the illuminated areas grow larger.

What was it like to be among the first readers of this luminous book? How many readers turned away from the candlelit pages and walked over to the window and stood gazing upward at the moon, seeing it anew? What was it like to be Galileo himself, sitting in the dark, propping the telescope up and peering through it and asking himself, "Can it be that I am actually watching sunrise on the moon?"

※

Anxious about his status as inventor and discoverer, Galileo once declared that his regard for the inventor of the harp was not reduced by his knowledge that the first one was a crude instrument poorly played. "To apply oneself to great inventions, starting from the smallest beginnings, is no task for ordinary minds; to divine the wonderful arts that lie hid behind trivial and childish things is a conception for super-human talents." His telescope was certainly crude by comparison with a 400-power computer-controlled toy available today in any shopping mall, but it was not the tool that was doing the observing; it was the mind behind the eye peering through it. The telescope merely channeled the moon's borrowed sunlight. Science needs the best possible tools and steadily refines them, and theorizing requires facts and must be subject to them. But not every eye peering through a telescope is capable of making discoveries that change our conception of the universe.

It is beyond ironic that the elderly Galileo went blind, one eye at a time, cutting short the observations and writing for his last book. Before he completely lost his sight, he again returned to his telescope. He trained his remaining feeble eye on the moon—and even then he saw what others had over-looked. He wrote to a friend that throughout history human beings had thought that the moon shows precisely the same face to Earth at all times: "I do not find that any change was ever noticed." But he had noticed a change around the edges of the visible circle of the moon's face. There was just enough movement to indicate that the side we see is not always pre-cisely aligned toward Earth, that there is a slight wobble that reveals a small amount more around the edges—a difference too slight to be noticed by the casual insomniac. Astronomers call this phenomenon the moon's libration, or rocking.

It was Galileo's last discovery. Soon he was completely blind. He lamented to a friend, accurately if not modestly, that the "universe which I with astonishing observations and clear demonstrations had enlarged a hundred, nay, a thousandfold beyond the limits commonly seen by wise men of all centuries past, is now for me so diminished and reduced, it has shrunk to the meagre confines of my body."

Sunrise on the Moon

Despite Galileo's apt and convincing analogies, sunrise on our natural satellite little resembles sunrise back on the home planet. Swaddled in dense atmosphere, Earth announces sunrise and sunset with free celestial light shows that can't be purchased at any price on the moon. Up there no clouds change hue at dawn, no bands of color define the horizon. The sun simply climbs up over the stark dry craters, casting those long jet-black shadows that caught Galileo's eye. However, it's not as if the sun just pops up over the horizon. A day on the moon—the time it takes it to rotate fully on its axis—is as long as a month on Earth: roughly thirty days. Naturally, at this constant pace, it also takes thirty times as long for the sun to rise, creeping up like a tethered balloon for an entire hour before it completely clears the lunar horizon. This day/month alignment is why the moon always shows us basically the same face (discounting the slight wobble around the edges that Galileo discovered), and also why lunar explorers never get to experience an Earthrise. The moon echoes Earth's relationship with its star by rotating on its own axis and revolving around its mother planet.

In fact, because our satellite is so large some astronomers describe the Earth/moon relationship as a two-planet system. As mentioned earlier when discussing the formation of Earth, astronomers reject the idea that our planet resulted from a collision between two astronomical bodies. Many experts, however, think that the collision theory may well explain the genesis of our oversize moon. Far larger in proportion to the planet it orbits than any other known satellite, the moon probably started out as fragments torn from Earth during a stupendous impact with another planet-size body. The debris would first have fallen into orbit around Earth and then gradually combined to form a permanent satellite.

After thousands of years of admiring and fantasizing about the moon's friendly old face, human beings didn't get to peek at its shy side until October 1959, when the Soviet Union achieved another first in space. Launched two years to the day after *Sputnik* became the first artificial satellite in space, and far more sophisticated than any previous craft, *Lunik 3* came within 40,000 miles (65,000 kilometers) of the dark side's surface. The camera system, which was unbelievably primitive by twenty-first-century standards, took forty minutes to process about thirty low-resolution photographs. Transmission back home took much longer. But the photos included enough of the sunlit edge of the moon to enable ecstatic Russian cartographers to begin naming areas: the Gulf of Cosmonauts, the Sea of Moscow, the Sea of Dreams. Of course, there is no actual "dark side of the moon," despite Pink Floyd's album; there is only the side that we do not see. Astronomers use the term "far side."

The Cup of the Immortal Fluid

In Hindu mythology the ferocious heat of the sun is contrasted with the life-giving generosity of the moon, which is thought to control not only tides but even the dew that seems to fall like manna from the night sky. These manifestations of water are the terrestrial analog of Amrita ("undead"), the ambrosialike drink of the gods. Such myths intuited a cycle of alchemical transformation that in time would be confirmed by science. Rain really does fall to the earth from the mysterious realm above us, and dew does indeed magically condense from the invisible air itself; plants quietly drink the vital fluid from the earth; cows eat plants that have nourished on the waters from heaven and transmute them into yet another sacred liquid, milk; human beings drink the milk and transform it into blood. To complete the circle, blood fuels the eyes and brain and imagination that perceive and honor the circle. While we gaze at the generous moon, our breath joins the rest of the unseen moisture in the air to eventually coalesce into rain that will begin the cycle all over again. Both mythology and science testify that it is impossible to draw a boundary of the self as if we were an independent entity. Circles have no beginning and no end. As the German anthropologist Heinrich Zimmer sums it up, "Amrita, water, sap, milk, and blood represent but differing states of the one elixir. The vessel or cup of this immortal fluid is the moon."

Eclipse

What is the moon? In one of the parenthetical asides that comprise half the novel, the narrator of Arno Schmidt's *Scenes from the Life of a Faun* nicely expresses how the human imagination both honors and tames the sovereign: "And as for me the sun is either a mere two fingers wide, or a belching golem of fire: all according to what use I have for it!" Schmidt is correct; even while driving eastward at sunrise, you can hold the flaming sun at bay with one upraised hand. It is odd to think that a 900,000-mile-wide fire looks so small in the sky, seems no bigger than the moon—or would, if we could reduce its glare and look directly at it. In fact, the two appear almost precisely the same size to us, at least in their actual diameter. Thanks to a most entertaining arrangement, the diameter of the sun is about four hundred times that of the moon, but the sun also happens to be about four hundred times as far away.

We forget this odd fluke of circumstance (a silly coincidence worthy of a bad science fiction novel) until an eclipse slides the moon over the sun for an almost perfect fit—an astronomical alignment that occurs every eighteen months or so, somewhere on Earth. It doesn't happen every month because the orbit of the moon is inclined a slight five degrees and nine minutes away from Earth's orbital plane, a situation requiring that the orbits be close to the node where they intersect before eclipse can occur. Any year will offer from two to five partial eclipses, but it is the total eclipse that dazzles us. How many planets have a moon that perfectly blots out the sun, leaving just enough room for a corona to glow around the edge? Imagine Blake's "pyramid of night" falling outward into space—with its umbra, the area of darkest shadow, only a small point of true darkness within the shadowy penumbra—and you can see why it rarely covers the moon precisely. (Incidentally, the Latin *umbra,* "shade," begat the Italian *ombrella* and led to that likable English word *umbrella.*)

It's possible, of course, for a planet's home star to be so far away that a satellite might obliterate its distant glimmer entirely, and technically this is still an eclipse. But those of us whose planet/satellite duo puts on such a regular show can be forgiven for sneering at other systems' feeble notion of eclipse. From Neptune, for example, the sun is so reduced in the black sky that any of its moons might completely block the star's light, but it wouldn't align in what we think of as an eclipse. Only on our own planet is the match between star and satellite so dramatic, so awe-inspiring. And surely this cinematic special effect is just as uncommon outside our local solar system.

No phenomenon better demonstrates the astronomical dance in which we are participating than an eclipse, either solar or lunar. Because the moon's orbit is elliptical, its distance from Earth is slightly different during each successive alignment. Its average distance is 239,000 miles (384,400 kilometers). However, at perigee—the point in a satellite's orbit when it comes nearest to whatever it is orbiting—it is 13,000 miles closer; during an eclipse at this time, its black silhouette entirely covers the sun. At apogee, however, when the moon is 252,000 miles away, it doesn't quite fill the entire circle of the solar glare. As a result, the sun's edge still shows around the perimeter of the moon in a brilliant corona. This version is called an annular eclipse, from the Latin word for "ring." (For the same reason, in Medieval Latin the third finger of the hand was called *annularis* because it was already established as the ring finger.)

A solar eclipse begins when the moon, until now invisible overhead and subjugated to the solar glare, touches the western edge of the sun. This is called first contact. Second contact occurs when the moon's eastern rim reaches the sun's eastern edge and covers the entire disk. The bright necklace of lights that you can sometimes see for an instant around the edge of the moon at this time results from sunlight bouncing around its mountainous horizon, glinting through uneven breaks in the edge of what seems from down here to be a perfect circle. This phenomenon is called Bailey's beads, after the astronomer who discovered its source.

If we are outdoors during a solar eclipse, we can experience—we are almost forced to experience—a reminder of the foreboding and alarm that such phenomena once inspired. Imagine that you do not live in the urban twenty-first century and you do not commute to work; you have not been hearing reminders of the approaching eclipse every half hour on TV news; you do not own a book in which the arrival of the next hundred solar and lunar eclipses is predicted to the very second. Imagine instead that you are an illiterate farmer in the Middle Ages, laboriously plowing a field behind a team of oxen on a hot summer day. Your attention is on the aches and pains you experience from being jerked along uneven furrows, and the number of recent plague deaths in your parish. Drenched with sweat, you would welcome shade but there is none to be had in this field. Then you realize that there seems to be a cloud covering the sun, and you glance upward, preparing to welcome a respite from the heat. But there is no cloud; there are none visible in the sky at all. Incredibly, something seems to be wrong with the sun. You halt your team and shade your eyes with a callused hand, and even as you watch the sun grows darker. The temperature is already going down, and the sweat on your back suddenly chills you. Other farmers in nearby fields have also halted their work and are staring upward at the sky. What is happening? The world is quickly growing darker. Within a few minutes, a rooster, who on a normal day would be silent for many more hours, begins to crow as if at dusk. Your oxen begin an uneasy lowing. You look down at the ground and see that flowers are beginning to close. Surely this kind of moment, this instant when the imagination confronts the mysteries of the natural world, is the genesis of mythology.

It is almost impossible for us to imagine the fear inspired by such ignorance of basic everyday realities behind human life. Eclipses were just as common then as now, but we hear about so many, of varying totality and in different parts of the world, that we forget that they are never common in a single area. They happen far enough apart that, in illiterate societies, they quickly became legendary. In our own time, every nature-oriented calendar includes details of solar and lunar eclipses, as easy to schedule as Mother's Day and Hal-loween. Unfortunately most of us witness this grand astronomical spectacle either through a window or at second hand, on a television screen.

One of the more tiresome aspects of nature study before, say, the secularization of the eighteenth and nineteenth cen-tury, was commentators' determination to force every natural phenomenon to bear the burden of a theological lesson. Yet the magnitude of such phenomena as eclipses reminds us that it is sometimes difficult to resist the urge to point a moral. Even if a solar eclipse rates only a minute at your office window, over your afternoon coffee, it still reminds us of how much larger the cosmos is than our narrow primate routine of feeding, sleeping, and worrying.

※

We are discussing eclipses this evening instead of earlier, dur-ing daylight, because the moon plays a leading role in both kinds of eclipse, solar and lunar. In the former it moves between us and the sun and becomes silhouetted; in the lat-ter we move between it and the sun and cast our shadow on it. What an odd thought, the vision of the shadow of our planet falling outward into (almost) empty space. Usually we don't think of a shadow as existing until it shows up on some object.

One such shadow is visible only after full darkness, and it is a much more common glimpse of celestial mechanics than a lunar eclipse. Often, under the right conditions, you can watch a satellite disappear into the shadow of the planet.

Many communication and survey satellites orbit Earth at roughly the speed of the planet's own rotation, which makes them appear to remain stationary above a certain part of the world. Such a geosynchronous (also called geostationary) orbit requires that a satellite move at a speed that matches the planet. So that it will orbit in twenty-four hours, a satellite needs to be placed at a height of 22,300 miles above the Earth. For comparison, a satellite placed 500 miles above Earth will require only a hundred minutes to circle the entire globe. (Incidentally, the idea for broadcasting over large areas of the globe with communication satellites placed in geostationary orbit was the brainchild of Arthur C. Clarke, of *2001* fame; he proposed such a system in 1945, two decades before *Intelsat I*, the first commercial satellite, became a reality.) One moment, as a satellite arcs over your head in the nighttime sky, it is glowing with the reflected light of the sun, and the next it has disappeared into Earth's shadow. Thousands of tons of metal objects now orbit over our heads, gradually coming down to burn up like a meteor in the atmosphere or, if large enough, to crash like an alien spaceship into the surface.

But of course no everyday occurrence, and nothing so visually subtle as a winking-out satellite, could ever equal an eclipse for drama. The American nature and science writer Jerry Dennis reported in one of his books how his children responded to the total lunar eclipse on August 16, 1989. Aaron was ten and Nick two. "Aaron," wrote Dennis, "saw animal shapes: a duck's head, a snapping turtle, a snake with its mouth wide open around an egg, a rock singer with a

Mohawk haircut, a Pac-Man character with its mouth frozen open." The much younger Nick, however, merely "watched in somber silence" until Earth's shadow completely covered the face of the moon, then announced in a sympathetic voice, "Moon's broke."

A child's compassion for the disappearing moon reminds us that the progress of Earth's shadow across the face of its satellite seemed miraculous to the authors of the Bible and other early works of mythology. That it turned out to be utterly predictable, not to mention far more reliable than your average fickle deity, in no way diminishes the crazy glory of the circumstance. And here's an entertaining historical side effect: the curved shadow of our planet, as it fell across the spotlit face in the sky, provided some of the earliest tangible evidence for a round rather than a flat Earth.

※

Tonight the moon, not broken, hangs whole and brilliant in the sky above our craning heads, casting dark shadows at our feet. At noon we considered shade and shadows with the sun as their source, but of course every light provides a shadow. It is possible to cast several at once on a typical urban night: abruptly running shadows from car lights, shadows that obediently heel for a block and then dodge around you as you pass between them and a streetlight. And—even in the light-blinding night of cities—you can see a faithful shadow that shortens as the moon climbs higher. The seventeenth-century Japanese poet Sodo observed his shadow walking home with him under a bright harvest moon. No doubt millions of other people have noticed the same phenomenon, but because he wrote down this experience Sodo and his shadow are still walking hundreds of years later, strolling along beside us and our own moon shadows.

Infinite

> *Earth cannot escape the sky.*
> —Meister Eckhart

The Lion or the Lobster?

The moon is gone, so now we can see the stars better. In our parade of celestial personages, you don't see the crowd behind the king until the royal chariot has passed.

Probably you have noticed during nighttime flights, as your plane banks in for a landing, how the streetlights below twinkle like the stars above, surrounding you with blinking constellations. From a plane you can perceive this fluctuation in the streetlights' illumination as soon as you are close enough to the ground to see street lamps at all. Their light flickers for the same reason as does the light from stars. It is passing through an atmosphere of varying warmth, an over-burdened ocean of air that transports vast quantities of suspended particles of what we dismiss collectively as dust—smoke and other pollutants, pollen, and microscopic amounts of flotsam that have drifted away from chemically unstable organisms such as ourselves. Light will bounce off of anything. The great Polish poet Wislawa Szymborska once remarked on how feeble a star's ray is, to be bent merely by bumping into space itself.

When you are near enough to the ground to distinctly see the lights and around them the streets and cars—and too close for photons to encounter enough intervening particles in the atmosphere to cause much twinkle—the streetlights may still flicker. The leaves of trees around them change position with every breeze to keep stirring as the plane flies overhead. The Christmasy twinkling effect is almost as beautiful in the square made world below as it is in the free sky above. If the plane's motion wasn't constantly changing our perspective, we might perceive the never quite random patterns of streetlights as constellation-worthy images: the Cloverleaf, the Mall, the River's Edge.

"The apportioning of stars into constellations is unnecessary," says the American novelist Nicholson Baker; "their anonymity enhances the sense of infinitude." Yes, but the brain's instinct for recognizing and imposing patterns cannot resist forming the stars into groups. A ladle, a scorpion, the cruciform outline of a flying swan—at least the key points of these images are discernible at a glance. But who were the drunken shepherds who first perceived a bird of paradise in the constellation Apus?

This morning we were interested in Ptolemy's geocentric cosmos, and how Copernicus and others later rejected it. As we gaze at the night sky, however, Ptolemy's other contributions come to mind. In the *Almagest*, published (not that "publication" then involved printing and widespread distribution) in the middle of the second century, he presented a list of forty-eight constellations that soon became the accepted guide to the northern night sky.

In a modern handbook to the sky, lines usually link the schematic white dots into a constellation, so that you perceive which distant suns officially participate in this group identity. Yet even a connect-the-dots diagram format isn't always enough to bridge the conceptual gap between the dots and lines and the image that they are supposed to inspire in your brain. Often you will also find a pale illustration behind the dots and lines. Otherwise how could you possibly know that the stars grouped into the constellation Corvus are supposed to represent a crow rather than a bent cardboard box? It is just possible to see the leaping dolphin in Delphinus if you're told to look for it, but landlubbers without star guides might see instead a kite on a string. Leo could just as well be the Sphinx or a sleeping camel or a mouse with its tail in the air. Recognizing the boundaries of constellations is the first hurdle for the amateur astronomer. Young students who have had little exposure to either mythology or astronomy might find both subjects more appealing if asked to reconfigure and rename constellations and then imagine stories as rationales for the new figures.

※

During the first half of the twentieth century, the German ethnologist Theodor Koch-Grünberg studied the native peoples of northern South America. In 1918 he published the first comprehensive study of the mythology of a South American people, the Arekuna of Venezuela and what is now

Guyana. One topic to which he returned with various groups was their perception of the night sky. Unaware of patterns that Europeans impose upon randomly distributed stars, when asked to draw constellations South American natives naturally portrayed quite different arrangements.

Koch-Grünberg asked individuals to draw how their people perceived the area around the constellation that Europeans call Leo, the Lion. He didn't specify which stars were included in the constellation. Where a constellation begins and ends, and which stars are included within its imaginary boundaries, is often debatable. Not surprisingly, in looking at what Koch-Grünberg thought of as Leo these equatorial coastal cultures did not perceive an awkward sideways picture of an animal that they had never observed. Instead they disregarded the stars that supposedly represent the lion's legs and saw an overhead view of a lobster. A member of the Mirití-tapuyo portrayed a lobster squared off like a broken trident. A Kobéua individual improved upon the constellation's meager allotment of stars and produced a detailed pointillist crustacean complete with grasping claws.

Skygazing seems to be a primordial urge built into our psyche. Not only does every part of the sky have a story; every part has *many* stories. Searching for the mythological patterns into which millennia of consensus has grouped stars is rather like seeking pictures in clouds, and it is an equally honorable and ancient pastime. It also has the virtue of providing neither moral instruction nor business opportunities (except for astrologers). Human beings have been looking up at the night sky since before we were human beings. Quite possibly grandfather *Australopithecus* surveyed the starry vault above Afar and grunted aloud at the omens. Our ancestors were plotting the motions of the sky long before they had systematically examined the world around them. As Joseph Wood Krutch, the American scholar and naturalist, observed, "The Greeks, who thought that bees were generated in the carcasses of dead animals and that swallows hibernated under the water, could predict eclipses, and the very Druids were concerned to mark the day on which the sun turned northward again."

The next step in the brain's predictable little dance is the creation of fictions to explain the patterns. Every constellation has a tall story behind it. Your punishment for drawing the ire of the gods might be metamorphosis into a cautionary fable visible for all the world to see. In mythology around the world, the gods always seem to be offhandedly turning somebody into a star or even into an entire constellation. To astronomers, the seven major stars—and their dimmer cohort—on the shoulder of the constellation Taurus are merely the open cluster Messier 45. To those Native Americans for whom stars represented the souls of the dead, the seven stars were the immortal essence of maidens who died on earth and were transported into the heavens—or sometimes only six, with the dimmer of the seven representing one who fell out of her place in the sky. In Greek mythology, from which we get the traditional Western name for this group, the stars are the Pleiades, the Seven Sisters, who fled

Orion and cried to Zeus for help. In the usual Greek mode of divine intervention showing up too late to be useful, Zeus turned the women into doves and placed them in the sky.

It is humbling to step outdoors and gaze up at what Czeslaw Milosz has called the inhabited classical sky. It turns out to be a dizzyingly busy place, inhabited by physics' best legerdemain and the daydreams of every generation that passed this way before we were born. How populous it is. According to Ovid, who wrote in the time of Augustus, even the emperor's immediate predecessor, Julius Caesar, now hangs in the sky, watching the scurry and fuss of human beings far below on the turning earth.

Inconstant Star

Of all the glittering lights in the sky, most of whose ancient stories we have forgotten, one is still acclaimed king of the heavens: the North Star. And Julius Caesar, of all people, shows up in its story too.

Even people who seldom glance at the night sky know that there is a star that somehow designates north. However, even the most random poll—of fellow occupants of an elevator, for example, or of your immediate neighbors at a cocktail party—will reveal that many people think the North Star is the brightest star in the night sky. They are also likely to assume that it is this star's impressive magnitude that makes it a reliable guiding light. For this reason they will march outdoors and confidently point to Vega or even to Mars.

You should challenge them to a bet, because they are mistaken. The star that we call Polaris is indeed a yellow super-giant, 50 times larger than the sun and broadcasting 2,500 times as much light—but it doesn't overwhelm the sky because it is 680 light-years away. This figure means that the North Star's photons touching your retinas at this moment left home about the time that John Wycliffe was laying the groundwork for Puritanism, and they have been racing inno-cently through space during all the rowdy history of the modern world. Because it is so far away, we assign Polaris a modest magnitude of 2.1, ranking it only among the bright-est couple of dozen stars visible from Earth. With all the competition it doesn't stand out much.

Therefore you must have help in locating it among the myriad. The first trick that an aspiring stargazer learns is to find the Big Dipper's bright and distinctive ladle shape, seven stars that form much of Ursa Major, the Big Bear. Then locate the two stars that form the side of the dipper farthest away from the curving handle, because they will serve as pointers to guide you to Polaris. Follow the angle of the pointer stars for about five times their distance apart, and you will find the North Star, as the end of the handle of the Little Dipper, which is also the tip of the tail of Ursa Minor, the Little Bear.

From our viewpoint down here on the surface of a spin-ning planet, this otherwise inconspicuous star appears to be the hub around which the rest of the sky rotates. Before the industrial revolution, when skies outside urban areas and over seas were still dark, the dutiful dance of the circumpolar constellations around Polaris was visible and crucial for ori-entation and navigation. Even in light-polluted urban areas, you can easily observe this phenomenon—and without sitting up all night and staring into space. A photographic exposure of only a few minutes will show the stars moving in arcs of light around a center point, and a long exposure will record them moving in full circles around the seemingly motionless Polaris.

It is such impressive constancy that makes the North Star both a guiding light for travelers and a popular figure of speech. But for armchair astronomers, literature's most famous invocation of this distant sun has surprising ramifications. "I am constant as the northern star," declaims Shakespeare's Julius Caesar, who probably would take time to summarize his noble virtues even if he knew that the Roman citizens gathered around him are heavily armed and planning his exit scene. And then the long-winded emperor elaborates on his simile, in case any barbarian in the crowd misses his point: "Of whose true-fix'd and resting quality / There is no fellow in the firmament." This scene takes place on the Ides (the fifteenth) of March in 44 BCE, the true date of Caesar's assassination. Shakespeare wrote it, however, in 1599. The gap in time between the inspiration and the reality tells a curious tale about the way the sky is changing. Should you ever be exposed to an outbreak of the recurring delusion that Francis Bacon wrote the works attributed to Shakespeare, you can refute even the most ardent Baconian (or at least provoke a nice argument) with a brief dissertation on the rotation of Earth. And now is the perfect time of day for it, as we stand under the evening sky and gaze at the stars overhead.

★

Francis Bacon was one of the most learned men of his time—a scholar of classical language and history, a founder of the Royal Society, and one of the first theorists of modern science, especially in his *Novum Organum*. He argued on behalf of evidence and experiment over classical tradition, so it is just conceivable that he might have appreciated a quirky evidence-based defense of Shakespeare. The man from Stratford was nowhere near as well schooled as Bacon; as Isaac Asimov and others have argued, he made a mistake in *Julius Caesar* that seems unlikely to have occurred to his illustrious contemporary. Bacon would never have placed in the mouth of Caesar the words "I am as constant as the northern star," because as a scholar of science he would have known something that Shakespeare did not—that in Caesar's time there *was* no northern star.

The North Star changes over time. We call it Polaris because it occupies a position at the North Celestial Pole, a point in the sky marked by an imaginary line running above the North Pole on Earth, as if extending its axis up from the surface, through the atmosphere, and out into space. There is also a South Celestial Pole, but no star serves as a South Star; only Sigma Octantis, so dim that without a telescope you can barely see it as a smudge of light, is even close. In fact, Polaris isn't precisely located at the center of those circling northern constellations, but it is within three-quarters of a degree, close enough to seem motionless amid its fellows. It wasn't very far from this position—only two degrees—four centuries ago, when Shakespeare was swilling ale and scribbling *Enter Caesar with attendants.* But in the first century BCE, five times further back in time, the earth's axis pointed at a different part of the sky.

How can this celestial position change over time? Usually astronomers explain that the stars appear to circle overhead because we are standing aboard a spinning top. This is such a familiar image for the rotation of the Earth on its axis that it has become a cliché, but it has the useful advantage in a metaphor of being apt in a secondary way. Like a top, Earth wobbles a bit on its axis as it spins, although we hope it isn't about to wind down and fall off the table. One consequence of this axial wobble is now called the precession of the equinoxes. Hipparchus, that brilliant noticer, realized what was going on back in the second century BCE, when he observed that the sun's position at sunrise inches gradually westward.

As we saw earlier today, with the story of Eratosthenes at noon, Earth's axis is tilted 23½ degrees away from the plane of the ecliptic, the plane in which it orbits the sun, and it is this skewed angle that provides us with the luxurious pleasure of seasonal variety. A perpendicular axis would prevent any such changes during the year. Over the last few decades astronomers have determined that the wobble is caused by the braking effect of the gravitational pull of both the sun and the moon on Earth's rotation, as the two forces struggle with each other. Although it doesn't change noticeably during a human lifetime, over the centuries the tilt moves. Naturally it also changes the orientation of everything else, not just Polaris, relative to us. The progress of this axial wobble is slow, however, and astronomers have calculated that it will take 25,800 years for it to move in a full circle. Halfway through the cycle from now, twelve thousand years distant, Earth's axis will point at another star that is visible over your head right now, Vega, the brightest in the constellation Lyra.

H. G. Wells's Time Traveller, who has been so obliging today in providing vivid imagery about time and its paradoxes, also helpfully supplies a scene in which, far in the future, he contemplates the sky above:

> Above me shone the stars, for the night was very clear. I felt a certain sense of friendly comfort in their twinkling. All the old constellations had gone from the sky, however: that slow movement which is imperceptible in a hundred human lifetimes, had long since rearranged them in unfamiliar groupings. . . . Through that long night I held my mind off the Morlocks as well as I could, and whiled away the time by trying to fancy I could find signs of the old constellations in the new confusion.

Incidentally, because of all this movement our reigning North Star, Polaris, will actually be at its closest alignment to the true Celestial North Pole around the year 2100. As mentioned above, right now it is three-quarters of a degree off. Polaris is moving ever closer to being the true North Star, at least for a while, but it will always be too inconstant to accurately reflect Julius Caesar's steadfast character.

Mean Time

Often during our journey today we have returned to the utterly taken-for-granted process that makes it possible—our planet's rotation on its tilted axis. From its first spin before it was a planet, through the nature of seasons and the reason for the Tropics of Cancer and Capricorn, to the reason why we have a North but no South Star, Earth's rotation has been the common theme. Therefore it is worth remembering here, while we gaze at the North Star, that twenty-first-century astronomers are able to monitor the precise angle of this influential tilt. They employ a sophisticated system of radio interferometry, a method of combining signals from linked radio telescopes that are located a long distance apart. Every year astronomers at about thirty-five locations around the world aim their radio telescopes at five hundred or so well-known and precisely located galaxies or quasars. Each telescope then times its receipt of radio waves, and their data are united into a picture of the Earth's orientation in space. The angle in which it has turned can be determined to within 0.0002 arcseconds, which in translation to ordinary human terms means within less than a centimeter.

What a useful antidote to hubris, to be reminded that our most basic assumptions about nature often turn out to be merely local traditions that have no meaning in the rest of the cosmos. Even our concepts of "day" and "year" are not fixed ratios; we wouldn't define them the same way on other planets. Earth's day (a single full spin on its axis) is a scant 1/365th of its year (a single full orbit of the sun), but the ratio differs on every other planet. Mercury, for example—the smallest planet in the solar system—can't just rotate on its axis and find the sun obediently repeating its journey across the sky. It moves too quickly in its revolution around the sun, because it is closer to it than any other planet. Indeed, this hyperactivity, visible over time even to the attentive naked eye, is why it was first named Mercury, after the fleet wing-footed messenger of the Roman gods. For the sun to move from one sunrise to another in the sky of barren Mercury, the planet must rotate three times on its axis. In the meantime it has orbited twice around the sun. This *spin-orbit coupling* results from millions of years of the sun exerting a tidal pull on Mercury, slowing its rate of rotation.

Earth's own rotation can be monitored without recourse to interferometry and networked telescopes. Astronomers can track the time at which particular stars cross the meridian at observatories around the world. In fact, this method was used to determine Greenwich mean time as late as the 1950s.

The Black of Night

Earlier today we journeyed with the Time Traveller himself, riding in the ornate sleigh that bore him into the future and back again to his own era. Ever since this pioneer story, time travel has been a favorite theme of science fiction. Physicists, many of whom acquired their interest in science as youthful fans of science fiction, speculate—because even equations sometimes have loopholes—about which arcane processes might permit travel within and around the confines of what we so casually call "time." Einstein theorized about variations in the experience of time, as in the famous twin paradox, in which a twin aboard a spaceship traveling near the speed of light would age considerably more slowly than the twin left behind on Earth. Experiments with atomic clocks have since proven him correct.

Archaeologists may come closer to traveling backward in time than the rest of us, when they are the first people in thousands of years to touch handmade artifacts in Teotihuacán or Abu Simbel. Surely it is a magical moment to unearth shards of baked clay that still preserve the shaping nudges of fingers and palms. To glimpse the past, the rest of us must be content with prowling family attics in search of evocative heirlooms, finding youth preserved in the scent of childhood books. The oddest of these records may be photographs of your parents happily frolicking in that vertiginous landscape that was the universe before you were born.

Nature preserves many glimpses of the past, of course, from mosquitoes fossilized in amber to the growth rings of the bristlecone pine. At Laetoli, Mary Leakey found preserved in hardened ash not only the footprints of two small pedestrians, but even molds of the raindrops that had turned the ash into a preserving medium. The most dizzying view we have of the past, however, comes not from the earth but from the sky, preserved not in stone or wood but in the most ephemeral medium of all—light. The night sky is a time machine.

Let's go backward in time by steps. Go to a natural history

museum. Stand under the articulate skeleton of a dinosaur—
Triceratops horridus, say, one of the most common of these
resurrected lizards. Aim your imagination like a spotlight to
bridge the chasm between this creature's arid bones and your
own wet flesh—its rib cage the size of a space capsule, the
twin horns, a huge curving plate of dragon armor atop its
head. Making this conceptual leap will not be easy. These are
mere bones, you will think, with your own pirate-flag skull
hidden by meat and busy with memories and worry. The
museum card will tell you that *Triceratops* lived roughly 65
million years ago. Imagine the humid air of a Cretaceous
morning as it sighs through those bony gaping nose-holes
and you will find your own nostrils dilating, suddenly detect-
ing the aftershave miasma emanating from a nearby member
of your own species. What did *Triceratops* smell on that long-
lost morning? What did it hear?

You have just imagined yourself 65 million years into the
past, so now you can do anything. Let's go further. Turn away
from the dinosaur. With barely a command from your brain,
your own secret bones will hold you upright and transport
you out through the front doors of the museum. Stand on the
steps and look up at the night sky. Thanks to the vast distances that their light has to travel before it reaches us, we
see stars as they looked years or centuries or eons ago. To say
that a star is 75 million light-years away is to say that we are
observing light that was flung from its source in the time of
the dinosaurs, before the asteroid impact that we discussed
at sunset. During all the subsequent evolution of life on
Earth, these photons have been traveling through almost

empty space. Think of the Time Traveller's first glimpse of unfamiliar constellations that had rearranged themselves over vast periods of time. So you need to stand under the stars to ask a question whose answer travels back in time to the beginning of our universe.

※

The natural sciences have the entertaining virtue of trying to provide all those answers we sought before, numbed by incurious conformity, we grew bored and forgot the questions. One of the most charming traits of small children is their insatiable yearning to know. The questions are endless: Why is the sky blue? How do birds fly? What is that thing over there? How does it work? Why? How do you know? These are very reasonable questions. Along the same line, there is a basic, simple question that astronomers have been asking for centuries, like children, and it is a wonderful question that never occurs to most of us: Why is the sky dark at night?

It isn't that we fail to understand what darkness is; darkness is simply the absence of light. Not that there haven't been other theories. There is the idea that darkness occurs when all the shadows come out from under trees and muddy the air, a charming image but not very helpful. And then there's the character in *The Third Policeman*, one of the outrageous novels by the Irish writer Flann O'Brien, who argues that darkness is actually a particular kind of black air that occurs only in the evening, the result of invisibly fine eruptions of volcanic material. In his theory, sleep is "a succession of fainting fits brought on by semi-asphyxiation." Let's ignore these fancies and take the question more seriously, and along the way we will encounter not only prominent astronomers but even that starry-eyed haunter of the nighttime, Edgar Allan Poe, proving to be a scientific detective worthy of his own C. Auguste Dupin.

Why, after all, should a universe populated with stars be dark? Think of Galileo Galilei in Padua, peering through his primitive telescope in the time of Shakespeare and chafing about ecclesiastical prohibitions against any theorizing that might actually be based upon evidence rather than classical ideals or religious dogma. After gazing in rapture at the divine magnitude of stars that had previously been invisible to him—so many times more than anyone had ever seen with the unassisted eye—Galileo wrote that through his telescope he perceived a host of stars "so numerous as to surpass belief." In his 1610 book *Sidereus Nuncius* (which we discussed earlier, while looking at the moon), Galileo maintained that the universe was an infinite space populated with an infinite number of evenly distributed stars. He sent a copy of this provocative book to Germany, to the astronomer Johannes Kepler, who had corresponded with Galileo before, when he sent the Italian a copy of his half-scientific and half-mystical volume *Mysterium Cosmographicum*. Barely a month later, Kepler replied that he had read *Sidereus Nuncius* and, among much praise and a few quibbles, disagreed with Galileo's contention about the infinite stars.

In his Renaissance observation-minded way, Kepler was quite the sun worshipper. Like Galileo, he followed the heretical thinking of Copernicus, who rejected the classical Ptolemaic view that Earth was the center of the cosmos and argued instead that the sun occupied that burdensome throne. But as a consequence, Kepler found himself horrified at the thought of the sun cast adrift among innumerable stars. To refute this midnight vision of chaos, he lobbied hard for a stable, finite cosmos centered on the reliable old sun, like a walled medieval city built around a castle. He insisted that surely, even to the most offhand observer, it was apparent that stars are not distributed uniformly throughout the universe. Also, because he was convinced that space is finite, he concluded that the magnitude of stars varies because they genuinely vary in size, not because they are spaced at differing distances from Earth. (It turns out that stars' magnitude varies for both reasons—size *and* distance.)

Kepler, with enterprising speed, soon turned his reply to Galileo into a pamphlet entitled *Conversation with the Starry Messenger*. He practices the time-honored art of disputational judo, using his opponent's momentum to throw him. Kepler argues that Galileo's contention that there are more than ten thousand visible stars, rather than supporting the Italian's case, actually strengthens Kepler's own argument *against* the infinity of the universe. Surely, he reasons, the combined magnitude of the stars would surpass the light of the sun. If Galileo is correct about the number and placement of stars in the cosmos, why doesn't their joined light blind us? Why is the night sky dark instead of glowing? Kepler insisted that the night sky is dark simply because there are too few stars in space to cover the entire sky with their light.

As physicists have pointed out ever since, Kepler missed a key point, which is that the size of an infinite number of stars is not a factor in their cumulative light if they are genuinely spread throughout an infinite cosmos. "Trees in a forest," wrote the American astronomer Edward Harrison in rebuttal to Kepler on this point, "form a continuous background no matter how thin their trunks." He provided another everyday example besides the now familiar trees-make-a-forest analogy: "The excited atoms in a candle flame emit pulses of light much too feeble to be detected individually by the eye. Their collective effect, however, accounts for the visible incandescence of the flame." But Kepler perceived correctly that an endless cosmos would mean an infinite amount of starlight, which is not what we experience.

⁂

Across the Channel and a century later, the English astronomer Edmund Halley carefully examined what he called the "metaphysical paradox" of the dark night sky. Now best remembered for recognizing the periodicity of the comet that still bears his name, Halley also performed many other noteworthy contributions to astronomy. By studying ancient and medieval records and comparing them with contemporary data, he became the first to discover that "some of the principal fixt stars" have changed position over time—a shocking revelation about the supposedly perfect and eternal heavens. Halley also discovered globular clusters, vast masses of ancient stars scattered throughout the cosmos.

More to our point here, he thought about the black of night. "In all these so vast spaces it should seem there is a perpetual uninterrupted day," he wrote, "which may furnish matter for speculation, as well to the curious naturalist as to the astronomer." Why then do we not perceive this light? Halley argued that, because light weakens over distance, such a vast expanse as it has to cross in space leaves tired light rays too faint to be detected by the human retina. Furthermore, as Halley pointed out, light's intensity decreases in proportion to the square of the distance involved.

Halley was correct when he argued that the light from any particular distant star might be too faint to broadcast its photons all the way to our eyes on Earth. But he was mistaken when he miscalculated the possible cumulative effect of all those glimmering stars out there. So astronomers kept coming back to a couple of key questions that Halley had hovered around but not quite grasped. First, is the universe too young for all of its light to have traversed its breadth in every direction? Or is it simply infinitely large and therefore harboring stars impossibly beyond even the mechanically aided limits of our perception? Many astronomers weighed in on these questions over the centuries—Thomas Digges in England, Jean-Philippe Loys de Chéseaux in France, and many others. Chéseaux, for example, proposed that not only is space not absolutely blank but in fact it has the capability of absorbing starlight.

But it was Heinrich Wilhelm Matthäus Olbers, a German, whose name has stuck to this quandary even though it was already well established when he was born in 1758. Besides being a part-time astronomer, he held down a successful and influential day job in Bremen as a physician, both a pioneer ophthalmologist and an early advocate of inoculation. Nights he peered at the sky and scribbled notes and became equally well-known for his astronomical accomplishments, including the discovery of the asteroids Vesta and Pallas and what long remained the standard method for calculating the orbits of comets.

Olbers proposed that space might not be infinite. "Are limits to it conceivable?" he asked. "And is it conceivable that the omnipotence of the Creator would have left this interminable space empty?" Reasoning from the-what-I-would-have-done-if-I-had-created-the-cosmos argument is quite common in the early history of astronomy (and not entirely unknown in the modern world), so we'll just keep moving without comment. Olbers suggested that not only was visible space sprinkled with suns "and their accompanying planets and comets," but that we will probably find the same to be true of the rest of the cosmos. Olbers complained that Halley had not sufficiently explained away the problem of the darkness of the night sky. Even distribution of stars throughout infinite space added up to an infinite number of stars, "for every line that we can imagine drawn from our eyes would necessarily lead to some fixed star, and therefore starlight, which is the same as sunlight, would reach us from every point of the sky." Stars would not only fill the sky;

they would be countless rows deep, each one hiding others behind it.

Then, as Chéseaux had done eight decades earlier—and Olbers possessed a copy of Chéseaux's work on this topic—Olbers proposed, without referring to his predecessor, that space was not simply a transparent blank. All of these theories so far assumed that space was a vacuum through which parallel rays of light moved unencumbered. "Now this absolute transparency of space," he adds in a dramatic tone of voice, "is not only undemonstrated but also highly improbable." And he went on, as Chéseaux had done before him, to suggest that perhaps the space between stars absorbed light. Olbers didn't cite Chéseaux, although historians argue that this action was likely an oversight rather than a fraud; there are many cases of people reading a source once and forgetting its argument over the years, only to find that details of it have surfaced in someone else's thinking.

But both were wrong. Even though this night sky quandary is now known as Olbers' Paradox, his theory too has failed the test of time. Astronomers have since determined that a truly bright-sky universe would require that temperatures in space equal those, as the astronomer Edward Harrison has remarked, "in a high-temperature furnace." This would result in a sky-wide blaze, "as if the Earth had plunged into the photosphere of the Sun." Over the two centuries after Olbers, other scientists joined the fracas and argued about the black of night. Fournier d'Albe in England suggested that perhaps the universe is populated more with dark stars than with bright ones, and that hot stars are actually uncommon, a situation that would result in an average cosmic temperature that he described as "quite comfortable."

✳

Again and again, the speculations of these intelligent and experienced men indicate that what would seem to be the simplest and most natural of questions turns out to be ever more complex and subtle. Then gradually the answer began to emerge, and from a very odd direction. In 1848, the year before his impoverished death in a Baltimore hotel, Edgar Allan Poe, always both fascinated and horrified by what he called "the abysmal distances" between "the multitudinous vistas of the stars," delivered a lecture entitled "The Cosmogony of the Universe." Some months later he published a revised version of his talk as the essay "Eureka: A Prose Poem." And it is here in Poe's offhandedly brilliant remarks, after all the many astronomers had had their say, that we find two parts of this great puzzle first coming together: the age of the cosmos and the speed of light.

"Were the succession of stars endless," wrote Poe,

then the background of the sky would present us an uniform luminosity, like that displayed by the Galaxy—*since there could be absolutely no point, in all that background, at which would not exist* a star. The only mode, therefore, in which, under such a state of affairs, we could comprehend the *voids* which our telescopes find in innumerable

directions, would be by supposing the distance of the invisible background so immense that no ray from it has yet been
able to reach us at all. [Poe's italics.]

When we look up at the night sky, we are seeing how stars
appeared in the distant past, when they emitted the light that
only now reaches our eyes. If stars are indeed distributed
relatively uniformly throughout the cosmos, then its edge,
the outer border of visible creation, would therefore be rushing away from us at the speed of light. Poe imagines that the
visible universe is comparatively small, with a much larger
not-yet-visible cosmos beyond. He speculates about stars'
light, "from the mere interval being so vast, that the electric tidings of their presence in Space, have not yet—through
the lapsing myriads of years—been enabled to traverse that
interval."

We are used to this concept nowadays—the vast age and
expanse of the universe—but neither was well established in
the public imagination in the time of Edgar Allan Poe. It is a
testament to the vigor of his imagination that it could caper
as elegantly in nonfiction as in poetry and fiction, because
scientific insight is always an act of imagination.

Astronomers continued to comment on this perennial
bugbear, of course, exploring the mechanisms behind the
shortage of light in the night sky. It is ironic that one of them,
the great English physicist Lord Kelvin, actually wrote about
this topic at length, and quite perceptively, but his paper was
omitted from his later bibliography and scholars lost it for
almost a century—the century during which this question

began to be known as Olbers' Paradox. Then, in 1912, Vesto Melvin Slipher, working at the Lowell Observatory in Arizona, began carefully analyzing the spectral lines of nebulae outside our galaxy to measure their distance from us. He was employing the now familiar phenomenon that astronomers call *redshift*. When the distance increases between an observer and an object that emits electromagnetic radiation, the radiation frequency decreases and its wavelength correspondingly increases. In a spectroscope, an instrument that channels a narrow beam of light through a prism or a diffraction grating to reveal its spectral fingerprint, the light from a receding star demonstrates an increase toward the red (longer-wavelength) end of the spectrum. By the 1940s it was established that the dimmer astronomical objects almost invariably exhibit the most prominent redshifts.

After a decade of painstaking work, Slipher proved that the great majority of the galaxies he observed showed a clear shift toward the red end of the spectrum. This shift indicated that the galaxies were moving away from us, much the same way that the pitch of a sound is higher as it approaches than as it recedes. This Doppler shift is demonstrated to you every time a siren passes, because such behavior applies to sound as well as to light.

Soon the astronomers Edwin Hubble (for whom the orbiting telescope would be named) and Milton Humason demonstrated that, far from simply fleeing us as if we were cosmically offensive, the galaxies out there are all moving away from each other. This utterly unforeseen and counterintuitive discovery—not predicted by any theory—was the final missing piece in the puzzle of the dark night sky. It demonstrated that the universe is expanding, which led to the notion of the Big Bang, and it explained why so much of the light out there has not yet reached our eager eyes. It took many scientists to reach this point, but standing proudly among them is Edgar Allan Poe, armed only with his wide reading and vivid prose and gamboling imagination, glimpsing the fascinating truth about the black of night.

Before dawn today we imagined ourselves back in history prior to the invention of artificial lights that kept darkness at bay. Later we went further back in time, tracing Earth's axial spin to its source in the gravitational forces forming the young solar system. But the story of the dark night sky takes us all the way back to the Big Bang, to the birth of this universe, to nature's genesis.

Names illuminate the sky like an aura.... It is we ourselves who give the stars their invisible reality, beyond the visible. By watching. By naming.

—Chet Raymo, *The Soul of the Night*

Swimming Allegories

The mind boggles not only before the distances and duration of the stars, but also before the myriad descriptions of stars that poets, novelists, lyricists, and naturalists have perpetrated. Let's avoid the wriggling catch of stellar metaphors and similes that you could net with a single scoop from the golden-oldies river of popular culture. We all know that both Vincent van Gogh and Don McLean love starry, starry nights. We all know that stars cross lovers and spangle banners, that they twinkle twinkle and fall on Alabama. Let's consider some less familiar responses to the many far suns that populate the night.

Whenever we are alone with the night sky or encounter fresh ideas about it, we find the stars sparkling anew. Then it is easier to understand how they planted themselves in the human imagination so early and so firmly. They are multitalented symbols. In the examples below, stars represent harmony, orderliness, precision, multiplicity, transcendence, and even metamorphosis. In a Passamaquoddy song from northeastern North America, the stars themselves proclaim that they sing with their light, because they are birds made of fire. Rabbit Angstrom, in John Updike's novel *Rabbit Is Rich*, envisions the countless dead hanging like stars in the black earth beneath his feet. An Ewe song from the eastern coast of Africa infers from the reliable circling of the stars that there are no accidents and no losses in nature, that "everything knows its way." The narrator of Nicholson Baker's novel *A Box of Matches*—neurotically attentive, as Baker's Nabokovian narrators tend to be—sees the predawn stars as "private needle-holes of exactitude in the stygian diorama."

Vladimir Nabokov himself repeatedly turned his characters' gaze to what he called in *Despair* "the swimming allegories of the starry vault," often with surprising responses from them. An unpredictable take on the night sky appears in his novel *The Real Life of Sebastian Knight*: "Years later Sebastian wrote that gazing at the stars gave him a sick and squeamish feeling, as for instance when you look at the bowels of a ripped-up beast." Nabokov echoes this horrific image in his later novel *Pnin*, during a scene in which three men stand gazing at the stars. One finally says, "And these are worlds," and another yawns and replies, "Or else a frightful mess. I suspect it is really a fluorescent corpse, and we are inside it." But it is in one of his short stories that Nabokov best expresses the sensation of insignificance that stars inspire: "People invent crimes, museums, games, only to escape from the unknown, from the vertiginous sky."

A glance at a single novel by one of Nabokov's favorite authors demonstrates that stars remain gainfully employed in literature because they are so adaptable. Stars show up as a favorite symbol in the work of Charles Dickens, and he saturated one novel with them. Their most obvious manifestation in *Great Expectations* is the name of Miss Havisham's ward, Estella, who seems—except perhaps in her unattainability—no more astral than, say, Stella in *A Streetcar Named Desire*. When Pip gets his first glimpse of the haughty Estella walking with a candle in hand, "her light came along the dark passage like a star," although hers proves to be devious light. Earlier, when Pip thinks of the felon cowering outdoors, he worries that a man might die while lying exposed on the marsh. Next Dickens expresses the compassionate but secular vision of life that underlies his pro forma pieties; he envisions at best a god like himself, distantly fond of his children but preoccupied elsewhere. "And then I looked at the stars, and considered how awful it would be for a man to turn his face up to them as he froze to death, and see no help or pity in all the glittering multitude." Later Biddy pauses in their nocturnal walk, looking at him "under the stars, with a clear and honest eye." The stars, the sky, God above—these trustworthy entities may observe Dickensian virtue, but decisions remain the burden of the individual trapped in life's muddle on the earth below. Pip expresses yet another vision of the witnessing heavens when he stands at the open kitchen door of the hut in which he was born and thinks, "The very stars to which I then raised my eyes, I am afraid I took to be but poor and humble stars for glittering on the rustic objects among which I had passed my life."

For millennia stars glimmered with significance in our actual nighttime. They guided camel caravans and sea voyages, clocked the passage of time, inspired mathematicians and poets. They became characters in our myths. We saw them every night and knew them as we now know the walls and ceiling of our homes. Most of us have lost this close personal relationship with the stars. We don't know how to navigate or tell time by them; we don't know their names or the stories behind the constellations into which our pattern-seeking brain automatically sorts them. But even in the twenty-first century, even paled to invisibility beyond our streetlights and exiled above our opaque ceilings, stars continue to glow in thought balloons over our preoccupied heads.

This Earth in Space

It's time to go indoors. As your mother often told you, you can't stay out here all night. As you enter your home, you see through the uncovered windows the stars visible above trees and buildings: the cosmos as background to your daily life. But when you turn on the light, all your windows instantly become mirrors that reflect the world you have created inside your home. The patient dark universe vanishes behind insistent artificial light and the clamor of furnishings. Familiar and cozy and beloved as this scene is, you pause at the door with your finger on the light switch, regretting the return indoors. There is your favorite chair, there a warm pool of golden lamplight, with a waiting book tented upside down on the coffee table. It's time for a cup of hot chocolate or perhaps a sip of scotch. But outside is the context for it all, the ancient stage behind this week's human comedy. You decide that you will read Thoreau or Annie Dillard and try to hold on to this fragile transcendent mood. After all, you don't have to see the stars to know that the cosmos is out there: vast, uncaring, populous, creative. Then you notice yourself reflected in the windows—*Interior with Figure*—and you realize that mirrors are fine for shaving or trying on a new dress, but facing the universe requires a see-through screen. Watching the reflected room, you turn out the light. Instantly you disappear. The nighttime mirror becomes a window again, and you can see the rest of the world and the stars beyond it.

Outside your house, beyond the wood and plaster, the

brick and particle board of your nest, the air moves with the turning earth. Oxygen swaddles our planet, but the blanket is thin, and only a few miles above your head it dissipates. Space is black and airless and does indeed make the green and blue splendor of Earth seem like the only flower blooming in a barren desert. Think of Yuri Gagarin tucked inside his claustrophobic spaceship in 1961, looking down at the way that Earth's blue sky fades slowly into the cold black of space. One of the recurring themes of early science fiction stories is the joyous response of astronauts as, after months or years out in the final frontier, they gaze again on the friendly blue-and-green globe of Earth, colorful and welcoming against the black void.

What a world, this Earth in space. Tilted, it points its magnetic pole at an otherwise insignificant star, and the constellations circle around it. A gigantic natural satellite, an arid stone parody of our wet planet, hangs in the sky, arcing across it slowly during the night, close enough to light us and far enough way to not threaten. Meanwhile sunlight falls on the other side of Earth, illuminating and warming. In one place the air is stirring, birds awaking; elsewhere bats greet the dusk. Slowly the line of demarcation between darkness and light creeps around the globe.

Obeying the commands built into the very rhythm of our planet, honoring the animal drives of our mammalian race who seek the comfort of nests and the warmth of fellows, we turn away from the daylight world and then from the night as well. We turn toward sleep—that great contradiction, the odd vulnerability of the resting animal, the mind's tantrum demand for recreation, for re-creation in metaphor every night. Our very metaphors have been molded on the turning globe, night after countless night, spun into our brains by eons of the world turning under the sleeping heads of bonobo and Neandertal. The pattern cannot be resisted; we must awaken and sleep; we must face both the day and the night. It's time for bed. The world will survive without your searching gaze for a few hours. One of the few things we know for certain is that, no matter what happens during the night, the sun will rise again.

Amber

There is a coda to the story of Phaethon, and like so many myths it is both beautiful and sad. When finally Clymene finds the bones of her son where the Hesperian nymphs buried them beside a riverbank, she falls upon the grave and weeps. Her three daughters join her, bruising themselves on Phaethon's marble tomb to cry for their lost brother, for his courage and folly. Four months they spend in lamentation. Then one day when the eldest sister, Phaethusa, starts to kneel upon the ground, she finds that she is unable to move her feet. She cries out. Beautiful Lampetie turns to race to her sister's aid, but finds that she too cannot move. She looks down, horrified, and sees her delicate young feet turning into gnarled roots that already grope for the soil beneath them. Distraught, the third sister clutches her own hair in agony— and pulls away handfuls of leaves. The girls feel bark armoring their shins; their arms lift of their own volition and form boughs, sprouting twigs and leaves. With a fiery itch, bark crawls across every inch of skin, leaving each of them only a pair of lips exposed.

With one voice they cry for their mother, who rises from her sorrowful collapse and turns from one daughter to the other in horror. She plucks at the bark as if to drag away an attacking animal, but it breaks off and tears the skin under-neath. The wounds bleed sap. The girls scream. Each cries out for her mother to stop, and just before the bark drowns her lips like a wave one of them manages to say, "Farewell!"

Eventually Clymene must give up her vigil. She has now lost all of her children. She abandons the graves of her son and the three saplings that had been his loving sisters and slowly she returns in sorrow to her homeland. The trees stand beside the river. They thrash in the wind. Now and then their wounds drip sap and gradually it hardens in the sunlight. In time pieces break off, fall to the ground, wash down the bank and into the water. The patient river carries the pebbles of hard yellow sap downstream to cities, where gem hunters find them and sell them to jewelers, who polish them and sell them to brides. The proud new owners call the jewels amber. They hold the lustrous gems up to the light to show how they glow and explain to jealous rivals that amber is a gift from the gods, distilled from Apollo's fire.

Terminat hora diem, terminat auctor opus.
"As the hour ends the day, the author ends his work."

—Christopher Marlowe, *Doctor Faustus*

ACKNOWLEDGMENTS

I could not possibly write a book about the day without being reminded how my wife, Laura Sloan Patterson, makes every day of my life more interesting and entertaining. With her usual insight and wit, Laura critiqued ideas and drafts as I worked on this book in Pennsylvania, North Carolina, Croatia, and England.

My excellent editor at Viking, Alessandra Lusardi, greatly improved *Apollo's Fire* by asking all the right questions, first in an overall critique of an early draft and then in a line-by-line edit of the final. Viking editor Molly Stern bought this book, as well as its predecessor, and helped shape its structure. I also want to thank production editor Barbara Campo, copy editor Roland Ottewell, jacket designer Jaya Miceli, book designer Carla Bolte, publicist Sonya Cheuse, proofreader Robert Legault, and indexer Laura Ogar. My agent, Heide Lange, has helped guide and secure my career for more than a decade. Thanks also to Heide's assistant, Alex Cannon; to the rest of the hardworking crew at Sanford J. Greenburger Associates, especially Teri Tobias in the foreign rights department; and to my British agent, Abner Stein, and his team at the Abner Stein Agency in London. The cordial staff of the Greensburg and Hempfield Area Library here at home are always helpful, especially Cesare Muccari, its director and my first friend in a new town; Diane Ciabattoni; and Cindy Dull. And thanks to the many libraries participating in the AccessPennsylvania system of interlibrary loans.

Astronomer and science historian William Sheehan discussed

many aspects of this book with me and critiqued numerous sections. Jonathan Miller discussed the original idea. Ross King and Matthew Chapman encouraged, critiqued, and supplied the luxurious pleasure of stimulating conversation; Ross also lobbied for the title. Many other people brainstormed, recommended sources, shared their expertise, or helped in other ways: Susan Abel, Denny Adcock, John Batteiger, Betty Jo and Harry Brown, Barb Ciampini, Paul and Valeria Cindric, Tamara Crabtree, Harvey Derrick, Michele Flynn and Collier Goodlett, Beecher Hedgecoth, Jim and Marty Hefner and the gang, Amy Garner Jerome and Aaron Jerome, Greg Norris, Robert Majcher, Laurie Parker, Bill and Rhonda Patterson, Sarah Patterson, Sally Schloss, Annie Sellick, Mark Wait, Dennis Wile, and F. Clark Williams. A lifetime merit badge goes to my brother and his wife, David and Jodi Sims.

In *Apollo's Fire* even more than in my earlier books, I want to thank three women in my family. First and foremost, thanks to my mother, Ruby Norris Sims, who made sure that my childhood preoccupation with the natural world was nourished with guides to stars and birds and dinosaurs, as well as a telescope, binoculars, and walks in the woods. My cousin Helen Derrick provided books, magazines, albums, and endless encouragement. And I want to remember here Mattie Hedgecoth, my great-aunt who died at the age of eighty-nine while I was writing this book. She ushered me into this world on a winter night in 1958, when the country doctor was delayed by a blizzard, and throughout my youth she was always next door, at the other end of the path through the woods, day and night.

SOURCES, NOTES, AND SUGGESTED FURTHER READING

The following notes provide sources for information and quotation, and also include works that explore topics in greater depth. If a listing omits publisher information, it is because the work is in the public domain and available in numerous editions. I do not cite every etymological or mythological reference because I consulted numerous sources for each.

Epigraphs

page xi: **"The first in time"**: Ralph Waldo Emerson, "The American Scholar," speech delivered 31 August 1837, in *Nature and Selected Essays*, ed. Larzer Ziff (London: Penguin, 2006).

page xi: **"The moon and sun"**: Bashō (Matsuo Munefusa), in *Narrow Road to the Interior*, translated by Sam Hamill (Boston: Shambhala, 1991).

page xi: **"The creation of the world"**: Marcel Proust, *The Fugitive*, Vol. 6 of *In Search of Lost Time*, translated by C. K. Scott Moncrieff and Terence Kilmartin, revised by D. J. Enright (New York: Random House, 1993).

Overture: Shadow Puppets on the Moon

page xxiii: **"for the most interesting and beautiful facts"**: Henry David Thoreau, *Journal*, entry for 18 February 1852.

page xxiii: **Coleridge . . . metaphors**: Richard Holmes, *Coleridge: Early Visions* (London: Hodder and Stoughton, 1989).

page xxiv: **"I have purposely dwelt"**: Charles Dickens, *Bleak House*, 1852–1853.

Morning Twilight

page 1: **"Gradually the dark sky"**: Dodie Smith, *I Capture the Castle* (New York: Little, Brown, 1949).

Stephen Dedalus's Cosmic Address

page 1: **Joyce's *Portrait***: James Joyce, *A Portrait of the Artist as a Young Man*, 1916.

page 2: **"The child transfigured"**: Dylan Thomas, "The Tree," *The Collected Stories of Dylan Thomas* (New York: New Directions, 1986).

page 4: **"We are nature seeing nature"**: Susan Griffin, *Woman and Nature: The Roaring Inside Her* (New York: HarperCollins, 1978).

The Threat of Dawn

page 7: **Borges wrote a poem**: Jorge Luis Borges, "Amanecer," translated by Stephen Kessler, in *Selected Poems*, ed. by Alexander Coleman (New York: Viking, 1999).

page 8: **Johnson famously replied**: James Boswell, *Life of Samuel Johnson*, 1791.

A Light in the East

page 10: **The atmosphere's effect on light rays**: James S. Trefil, *The Unexpected Vista: A Physicist's View of Nature* (New York: Scribner's, 1983) and *Meditations at Sunset: A Scientist Looks at the Sky* (New York: Scribner's, 1987); Hans Christian von Baeyer, *Rainbows, Snowflakes, and Quarks: Physics and the World Around Us* (New York: Random House, 1984); Carl Sagan, *Pale Blue Dot: A Vision of the Human Future in Space* (New York: Random House, 1994).

page 12: **"We might just as reasonably"**: Sagan, *Pale Blue Dot*.

page 14: **"The Earth has a very"**: Yuri Gagarin, quoted in Sagan, *Pale Blue Dot*.

page 14: **Francisco Pacheco**: Alberto Manguel, *Reading Pictures* (New York: Random House, 2001); Zahira Veliz, ed., *Artists' Techniques in Golden Age Spain: Six Treatises in Translation* (Cambridge, UK: Cambridge University Press, 1986).

page 15: **Fill a fishbowl:** adapted from Hans Christian von Baeyer and tested by the author.

page 16: **Newton demonstrated**: James Gleick, *Isaac Newton* (New York: Pantheon, 2003); Carl B. Boyer, *The Rainbow: From Myth to Mathematics* (New York: Thomas Yoseloff, 1959).

page 18: **"I say, the sun"**: George Gordon, Lord Byron, *Don Juan*, Canto the Second. This canto was first published in 1818–1819.

Smiley

page 19: **deity of ritual and fire**: Ajit Mookerjee, *Ritual Art of India* (London: Thames & Hudson, 1998).

page 19: **Dutch artist emulated . . . Hokusai and other**: Nicholas Wadley, *The Drawings of van Gogh* (London: Paul Hamlyn, 1969); Robert Wallace, *The World of van Gogh* (New York: Time Inc., 1969); Gabriele Fahr-Becker, *Japanese Prints* (New York: Taschen, 2000).

page 20: **"I used to be amused"**: Nicholson Baker, *A Box of Matches* (New York: Random House, 2003).

A Perfect Story

page 20: **"Every day . . . is an artistic whole"**: Christopher Morley, "Between Two Chapters," in *Essays* (Garden City, NY: Doubleday, Doran, 1928).

page 21: **the titular Mr. Phillips:** John Lanchester, *Mr. Phillips* (New York: Putnam, 2000).

page 21: **while she composed her novel**: Virginia Woolf, *Mrs. Dalloway* (London: Hogarth, 1925); Woolf, *The Diary 1920–1924*, ed. Andrew McNeillie (London: Hogarth, 1978); Francine Prose, ed., *The Mrs. Dalloway Reader* (New York: Harcourt, 2003); Quentin Bell, *Virginia Woolf: A Biography* (New York: Harcourt, 1972).

page 22: **"The sun sharpened"**: Virginia Woolf, *The Waves* (London: Hogarth, 1931).

page 22: **Updike observed**: John Updike, "2000, Here We Come," in *More Matter* (New York: Knopf, 1999).

Chariot

The Endless Array

page 23: **"Morning has broken"**: Eleanor Farjeon, "Morning Has Broken," from *Songs of Praise*, ed. by Percy Dearmer et al. (Oxford: Oxford University Press, 1931).

page 23: **"Dawn ignites"**: Meng Hao-jan, *The Mountain Poems of Meng Hao-jan*, translated by David Hinton (New York: Archipelago, 2006).

page 25: **"It was five o'clock"**: Iris Murdoch, *The Sacred and Profane Love Machine* (New York: Viking, 1974).

page 26: **Lyell ... Eocene:** Stephen Jay Gould, *Time's Arrow, Time's Cycle: Myth and Metaphor in the Discovery of Geological Time* (Cambridge, MA: Harvard University Press, 1988).

page 26: **Thomas Huxley named:** Adrian Desmond, *Huxley: From Devil's Disciple to Evolution's High Priest* (Reading, MA: Addison-Wesley, 1997); Richard Milner, *The Encyclopedia of Evolution* (New York: Henry Holt, 1993).

page 27: **"Only that day dawns":** Henry David Thoreau, *Walden* (Boston: Ticknor and Fields, 1854).

page 28: **"It was on that road":** Colette, *Sido*, translated by Enid McLeod (New York: Farrar, Straus and Giroux, 2002).

page 28: **Baudelaire did not respond:** Charles Baudelaire, *Les Fleurs du Mal*, translated by William Aggeler (Fresno, CA: Academy Library Guild, 1954). For a provocative take on this poem's day/light imagery (although using a different translation), see Eliane DalMolin, *Cutting the Body: Representing Woman in Baudelaire's Poetry, Truffaut's Cinema, and Freud's Psychoanalysis* (Ann Arbor: University of Michigan Press, 2000).

page 29: **"For the rest of my life":** Albert Einstein, according to fellow physicist Wolfgang Pauli; quoted in Ronald W. Clark, *Einstein: The Life and Times* (New York: World, 1971).

A Year of Light

page 29: **"Science shows us":** Heinz R. Pagels, *The Cosmic Code: Quantum Physics as the Language of Nature* (New York: Simon & Schuster, 1982).

page 30: **Gamow ... Einstein ... Planck:** Sidney Perkowitz, *Empire of Light: A History of Discovery in Science and Art* (New York: Henry Holt, 1996); David Bodanis, *E=mc²: A Biography of the World's Most Famous Equation* (New York: Walker, 2000); Michael White and John Gribbin, *Einstein: A Life in Science* (New York: Dutton, 1994).

page 32: **Mark Twain once:** Albert Bigelow Paine, ed., *Mark Twain's Letters*, 2 vols. (New York: Harper & Row, 1917); see also Charles Neider's comments on Paine's remark in Neider's introduction to his *Selected Letters of Mark Twain* (New York: Cooper Square Press, 1989).

Down to Earth

page 34: **laws of planetary motion:** Kitty Ferguson, *Tycho and Kepler: The Unlikely Partnership That Forever Changed Our Understanding of* *the Heavens* (New York: Walker, 2002); Owen Gingerich, *The Eye of Heaven: Ptolemy, Copernicus, Kepler* (Melville, NY: American Institute of Physics Press, 1993); Margaret Wertheim, *The Pearly Gates of Cyberspace: A History of Space from Dante to the Internet* (New York: Norton, 1999).

page 35: "**Yesternight the sun**": John Donne, "Song: Sweetest love, I do not go," 1611.

Sun Sister and Moon Brother

page 36: **myth of the Yuchi . . . Raven of Kwakiutl**: John Bierhorst, *The Mythology of North America* (Oxford: Oxford University Press, 2002); David M. Jones and Brian L. Molyneux, *Mythology of the American Nations* (London: Hermes House, 2004). For more on the Kwakiutl, see the first note to "The Law of Dance" in "Dance."

page 37: **dual symbolic classification**: Rodney Needham, ed., *Right and Left: Essays on Dual Symbolic Classification* (Chicago: University of Chicago Press, 1974); Alan W. Watts, *The Two Hands of God: The Myths of Polarity* (New York: Braziller, 1970).

page 38: **syncretism**: Frederick C. Grant, *Hellenistic Religions: The Age of Syncretism* (London: Macmillan, 1953); for a contemporary take on this idea, see Roy Ascott, "Syncretic Reality: Art, Process, and Potentiality," in *Drain* (online at www.drainmag.com), 2005; Wendy Doniger, ed., *Merriam-Webster's Encyclopedia of World Religions* (Springfield, MA: Merriam-Webster, 1999).

page 38: "**front and back follow**": Lao Tsu, *Tao Te Ching*, translated by Gia-Fu Feng and Jane English (New York: Vintage, 1972).

Spin

pages 39–40: "**When we try to pick out**": John Muir, *My First Summer in the Sierras* (Boston: Houghton Mifflin, 1911).

page 41: **composed the primordial Earth**: Ian Ridpath, ed., *The Illustrated Encyclopedia of the Universe* (New York: Watson-Guptill, 2001); Arthur Upgren, *Many Skies: Alternative Histories of the Sun, Moon, Planets, and Stars* (New Brunswick, NJ: Rutgers University Press, 2005).

page 42: **poisonous clouds of Venus**: Upgren, *Many Skies*.

Apollo Bright and Pure

page 43: **proud daylight than Apollo**: Michael Grant, *Myths of the Greeks and Romans* (New York: World, 1962); Robert Graves, *The Greek Myths* (London: Penguin, 1988); Roy Willis, ed., *World Mythology* (New York: Henry Holt, 1993); Tamra Andrews, *A Dictionary of Nature Myths: Legends of the Earth, Sea, and Sky* (Oxford: Oxford University Press, 2000).

page 45: **"Krypton's last son"**: *Action Comics* #14, July 1939; Lois H. Gresh and Robert Weinberg, *The Science of Superheroes* (New York: Wiley, 2002). The character of Superman was created by Jerry Siegel and Joe Shuster in 1938.

page 45: **The Egyptians credited**: Alfred Wiedemann, *Religion of the Ancient Egyptians* (London: H. Grevel, 1897).

page 45: **gods were portrayed with long hair**: For more on this topic and other mythological associations between the body and astronomical phenomena, see my book *Adam's Navel: A Natural and Cultural History of the Human Form* (New York: Viking, 2003), and the bibliography therein.

The Son of Fire and Water

page 46: **Phaethon** (throughout the book): For the narrative in my remake of Ovid's Phaethon tale, from Book II of his *Metamorphoses*, I relied upon four translations: Samuel Garth, John Dryden, et al., 1717; Allen Mandelbaum, *The Metamorphoses of Ovid* (New York: Harcourt Brace, 1993); Rolfe Humphries, *Metamorphoses* (Bloomington: Indiana University Press, 1955); A. D. Melville, *Metamorphoses* (Oxford: Oxford University Press, 1998). The words of this shorter version, however, are my own; for my purposes, I conflate the characters (or at least the names) Phoebus, Helios, and Apollo. See also Philip Hardie, ed., *The Cambridge Companion to Ovid* (Cambridge: Cambridge University Press, 2002).

page 49: **enrich Byron's *Don Juan***: "the god / Had told his son to satisfy this craving / With the York mail," from *Don Juan*, Canto the Tenth. This canto was first published in 1823.

Chariot

page 50: **recurring imagery about the chariot of the sun**: Walter Torbrugg, *Prehistoric European Art* (New York: Harry N. Abrams, 1968); Ethel L. Urlin, *Festivals, Holy Days, and Saints' Days: A Study in Origins and Survivals in Church Ceremonies and Secular Customs* (London: Simpkin, Marshall, Hamilton, Kent and Co., n.d.; reprint, Detroit: Gale Research Co., 1979); J. E. Cirlot, *A Dictionary of Symbols* (London: Routledge and Kegan Paul, 1971); J. C. Cooper, *An Illustrated Encyclopedia of Traditional Symbols* (London: Thames & Hudson, 1978).

page 51: **Krishna's disk ... *cakra***: Heinrich Zimmer, *Myths and Symbols in Indian Art and Civilization* (Princeton, NJ: Princeton University Press, 1972).

Dance

Day of the Sun

page 55: **"A strict law"**: This Kwakwaka'wakw axiom is quoted in many sources without attribution, including in the documentary film *Potlatch: A Strict Law Bids Us Dance*, directed by Dennis Wheeler, 1974. *Kwakwaka'wakw* means "people who speak Kwak'wala." Formerly this First Nations people of western Canada, along with several others, were called Kwakiutl, but this term is now restricted to a single group on Vancouver Island.

The Law of Dance

page 57: **Einstein demonstrated**: Michael White and John Gribbin, *Einstein: A Life in Science* (New York: Dutton, 1994); Ronald W. Clark, *Einstein: The Life and Times* (New York: World, 1971).

page 58: **Ptolemy ... Copernicus**: Owen Gingerich, *The Eye of Heaven: Ptolemy, Copernicus, Kepler* (Melville, NY: American Institute of Physics Press, 1993); Kitty Ferguson, *Measuring the Universe: Our Historic Quest to Chart the Horizons of Space and Time* (New York: Walker, 1999); A. R. Hall, *The Scientific Revolution, 1500–1800: The Formation of the Modern Scientific Attitude* (Boston: Beacon, 1954); Herbert Butterfield, *The Origins of Modern Science* (New York: Free Press, rev. ed., 1965); Stephen Toulmin and June Goodfield, *The Fabric of the Heavens: The Development of Astronomy and Dynamics* (New York: Harper and Bros., 1961).

Not Invisible

page 63: **Leonardo da Vinci distinguished**: *The Notebooks of Leonardo da Vinci*, translated and edited by Edward MacCurdy (New York: Braziller, 1955); Michael White, *Leonardo: The First Scientist* (New York: St. Martin's, 2000); Serge Bramly, *Leonardo: Discovering the Life of*

Leonardo da Vinci, translated by Siân Reynolds (New York: Harper-Collins, 1991).

page 63: **"The impact of the appearance"** and all other da Vinci quotations: MacCurdy, ed., *Notebooks*.

page 66: **"The air is not invisible"**: Hans Christian von Baeyer, *Rainbows, Snowflakes, and Quarks: Physics and the World Around Us* (New York: Random House, 1984).

Mere Fire

page 68: **"A stone look"**: Richard Wilbur, "Advice to a Prophet," in *The Poems of Richard Wilbur* (New York: Harvest, 1963).

page 70: **"We are receiving"**: Henry David Thoreau, *Journal,* entry for 7 September 1851.

High Noon

page 72: **"I did so love my body"**: Don Marquis, "A Ghost Speaks," in *Poems and Portraits* (Garden City, NY: Doubleday, Page, 1922).

Shadow Show

page 73: **a little boy named Milo**: Norton Juster, *The Phantom Tollbooth* (New York: Knopf, 1961).

page 76: **Norman Rockwell**: Norman Rockwell, *My Adventures as an Illustrator* (Garden City, NY: Doubleday, 1960).

Hunt the Shadow

page 79: **The sundial**: Gerard L'E Turner, *Scientific Instruments, 1500–1900: An Introduction* (London: Philip Wilson, 1998); J. B. Priestley, *Man and Time* (London: Aldus, 1964); Samuel A. Goudsmit, Robert Claiborne, et al., *Time* (New York: Time Inc., 1966).

page 80: **"He first put a sundial"**: Pliny the Elder, *Natural History: A Selection*, translated and edited by John F. Healy (London: Penguin, 1991).

page 82: **Thomas Jefferson liked**: Silvio A. Bedini, *Thomas Jefferson: Statesman of Science* (New York: Macmillan, 1990).

Measure the Earth

page 83: **Eratosthenes**: E. Gulbekian, "The Origin and Value of the Stadion Unit Used by Eratosthenes in the Third Century B.C.," *Archive for*

History of Exact Sciences 37:4 (1987); Giorgio de Santillana, *The Origins of Scientific Thought: From Anaximander to Proclus, 600 B.C. to A.D. 500* (Chicago: University of Chicago Press, 1961); Carl Sagan, *Cosmos* (New York: Random House, 1980); Chet Raymo, *Walking Zero: Discovering Cosmic Space and Time Along the Prime Meridian* (New York: Walker, 2006).

page 88: **"Eratosthenes' drawing"**: Raymo, *Walking Zero.*

The Equation of Time

page 88: **Edward Emerson Barnard**: William Sheehan, *The Immortal Fire Within: The Life and Work of Edward Emerson Barnard* (Cambridge, UK: Cambridge University Press, 1995).

page 89: **solar enlarging camera**: Sheehan, *Immortal Fire*; Jack Wilgus, "Professor Woodward: Pioneer Photographic Inventor and Educator," in *Forays* (Maryland Institute College of Art, 1996).

page 90: **"Through summer's heat"** and other Barnard quotations: quoted in Sheehan, *Immortal Fire,* from an autobiographical sketch in the Barnard papers in Vanderbilt University Special Collections and Archives.

page 91: **"I wield the flail"**: Percy Bysshe Shelley, "The Cloud," 1820 (published with *Prometheus Unbound*).

Written on the Sky

page 92: **Clouds**: Gavin Pretor-Pinney, *The Cloud-Spotter's Guide: The Science, History, and Culture of Clouds* (New York: Perigee, 2006).

page 94: **Luke Howard**: Richard Hamblyn, *The Invention of Clouds: How an Amateur Meteorologist Forged the Language of the Skies* (New York: Farrar, Straus and Giroux, 2001).

Artificial Clouds

page 99: **contrails**: "Hot Trails: To Fight Global Warming, Kiss the Red-eye Good-bye," by Christina Reed, *Scientific American*, September 2006; Pretor-Pinney, *The Cloud-Spotter's Guide.*

Gunfight in Abilene

page 103: ***High Noon***: Fred Zinnemann, director, and David Duncan, screenwriter, *High Noon* (United Artists, 1952).

page 104: **telegraph**: John R. Stilgoe, *Outside Lies Magic: Regaining*

History and Awareness in Everyday Places (New York: Walker, 1998); Tom Standage, *The Victorian Internet: The Remarkable Story of the Telegraph and the Nineteenth Century's On-line Pioneers* (New York: Walker, 1998).

page 104: **"And around noon"**: Stilgoe, *Outside Lies Magic*.

Firmament

page 106: **"The sky constitutes"**: Chet Raymo, *Natural Prayers* (St. Paul, MN: Hungry Mind, 1999).

The Flapping of a Black Wing

page 109: ***The Time Machine***: book, H. G. Wells, 1895; film, George Pal, director, and David Duncan, screenwriter (Metro-Goldwyn-Mayer, 1960).

A Strong Metal Bowl

page 110: **Writing about torture**: John Berger, "The Hour of Poetry," from *The Sense of Sight* (New York: Pantheon, 1986).

page 112: **Sky Woman**: T. C. McLuhan, ed., *The Way of the Earth: Encounters with Nature in Ancient and Contemporary Thought* (New York: Touchstone, 1994).

page 112: **Milosz called "the inhabited"**: Czeslaw Milosz, from the poem "A Family," in *New and Collected Poems, 1931–2001* (London: Allen Lane, 2001). No translator identified.

page 113: **The English word *firmament***: Paul J. Achtemeier, ed., *The HarperCollins Bible Dictionary* (San Francisco: HarperSanFrancisco, 1996); Mircea Eliade, *The Sacred and the Profane: The Nature of Religion*, translated by Willard R. Trask (New York: Harcourt, 1959); Isaac Asimov, *The Left Hand of the Electron* (Garden City, NY: Doubleday, 1972).

page 115: **the Fourth Sun of creation**: C. Scott Littleton, ed., *Mythology: The Illustrated Anthology of World Myth and Storytelling* (London: Duncan Baird, 2002).

page 115: **The Kadaru and Niyamang peoples**: Susan Feldmann, ed., *African Myths and Tales* (New York: Dell, 1963); Jan Knappert, *African Mythology* (London: Diamon, 1995).

page 115: **"As I walked down"**: Ken Bruen, *The Guards* (New York: St. Martin's, 2003).

A White-Hot Arrow

page 117: "**too—darn—hot**": Cole Porter, "Too Darn Hot," from *Kiss Me Kate*, 1948.

page 118: "**Go down, old Hannah**": from "I Vision God," in Zora Neale Hurston, *Jonah's Gourd Vine* (Philadelphia: Lippincott, 1934).

What Makes the Wind Blow?

page 120: **variations in temperature and humidity**: Jan DeBlieu, *Wind: How the Flow of Air Has Shaped Life, Myth, and the Land* (Boston: Houghton Mifflin, 1998); Scott Huler, *Defining the Wind: The Beaufort Scale, and How a 19th-Century Admiral Turned Science into Poetry* (New York: Crown, 2004).

Policeman Ozone

page 123: **the strange story of ozone**: J. R. McNeill, *Something New Under the Sun: An Environmental History of the Twentieth-Century World* (New York: W. W. Norton, 2000); John Emsley, *Molecules at an Exhibition: The Science of Everyday Life* (Oxford: Oxford University Press, 1999); Arthur Upgren and Jurgen Stock, *Weather: How It Works and Why It Matters* (Cambridge, MA: Perseus, 2000); Gale E. Christianson, *Greenhouse: The 200-Year History of Global Warming* (New York: Walker, 1999).

page 129: "**had more impact on the atmosphere**": McNeill, *Something New Under the Sun*.

A Presumptuous Smoke

page 131: **Evelyn wrote a pamphlet**: Charles Officer and Jake Page, *Tales of the Earth: Paroxysms and Perturbations of the Blue Planet* (Oxford: Oxford University Press, 1993).

page 131: **it took a disaster**: McNeill, *Something New Under the Sun*.

The Two Most Beautiful Words

page 137: "**The sweat that drenched him**": Leo Tolstoy, *Anna Karenina*, translated by Richard Pevear and Larissa Volokhonsky (New York: Viking, 2001).

page 138: **Wallace Stevens's famous poem**: "A Rabbit as King of the Ghosts," in *The Collected Poems of Wallace Stevens* (New York: Knopf, 1954).

page 138: "**Letting the days**": David Byrne, "Once in a Lifetime," performed by Talking Heads on *Remain in Light* (Warner Bros., 1980).

page 139: " '**most beautiful words**' ": Edith Wharton, *A Backward Glance* (New York: Scribner's, 1934).

page 139: **nearing brillig**: Lewis Carroll, *Through the Looking-Glass, and What Alice Found There*, 1871.

page 139: *Meles helicosaurus*: coined by the author, based upon Carroll's description, with a genus name borrowed from European badgers.

page 139: **in an 1855 pamphlet**: Morton N. Cohen, *Lewis Carroll: A Biography* (New York: Knopf, 1995); *The Annotated Alice: The Definitive Edition,* ed. Martin Gardner (New York: Norton, 2000).

The Silent Concert

The Path of the Light

page 141: "**When the sun begins**": Italo Calvino, *Mr. Palomar*, translated by William Weaver (New York: Harcourt, 1986).

page 142: "**The sun was now gone**": Nathaniel Hawthorne, "The Snow Image," from *The Snow Image and Other Twice-Told Tales*, 1852.

page 143: **Rulfo's novel**: Juan Rulfo, *Pedro Páramo*, translated by Margaret Sayers Peden (New York: Grove, 1994).

page 143: **Schmidt's *Scenes***: Arno Schmidt, *Scenes from the Life of a Faun*, translated by John E. Woods (New York: Marion Boyars, 1983).

page 143: "**Today I saw a red-and-yellow sunset**": Woody Allen, "Selections from the Allen Notebooks," in *Without Feathers* (New York: Random House, 1975).

Diamond Dust and Sun Dogs

page 146: *diamond dust . . . sun dog*: Richard A. Keen, *Skywatch East: A Weather Guide* (Golden, CO: Fulcrum, 1992).

Sunset Races

page 149: "**We go eastward**": Henry David Thoreau, "Walking," published posthumously in 1862 in the *Atlantic Monthly* but begun in 1851.

page 149: "**To Americans I hardly**": The first line of this collaborative couplet is by Thoreau and the second ("Westward the star") he slightly misquoted from the eighteenth-century Irish bishop George Berkeley,

"On the Prospect of Planting Arts and Learning in America," in which the subject was "course" rather than "star." Lewis Hyde suggests that instead of misquoting Berkeley, Thoreau may have been accurately quoting John Adams, who in his 1802 "Oration at Plymouth" cites Berkeley but misquotes in the same way. See Lewis Hyde, ed., *The Essays of Henry D. Thoreau* (New York: North Point/FSG, 2002). This is the same Berkeley referred to in "The Threat of Dawn."

page 149: **Alkínoös introduces Odysseus**: Homer, *The Odyssey*, translated by Robert Fitzgerald (New York: Doubleday, 1961).

page 150: **"no doubt that the Ethiopians"**: Pliny the Elder, *Natural History: A Selection*, translated and edited by John F. Healy (London: Penguin, 1991).

All Those Cares and Fears

pages 153–154: **Mount Pinatubo . . . Tambora . . . "the year without a summer"**: See various entries in David Ritchie and Alexander E. Gates, *Encyclopedia of Earthquakes and Volcanoes* (New York: Facts on File, 2001); Arthur Upgren and Jurgen Stock, *Weather: How It Works and Why It Matters* (Cambridge, MA: Perseus, 2000); Richard A. Keen, *Skywatch East: A Weather Guide* (Golden, CO: Fulcrum, 1992).

page 154: **four English tourists**: Emily W. Sunstein, *Mary Shelley: Romance and Reality* (Boston: Little, Brown, 1989); Muriel Spark, *Mary Shelley* (London: Penguin, 2002).

page 155: **"The wind, which had hitherto"**: Mary Shelley, *Frankenstein, or, The Modern Prometheus*, published anonymously in 1818 and under her own name in 1831.

Sic Transit

page 155: **the colossal impact of the meteorite**: Stephen Jay Gould, ed., *The Book of Life: An Illustrated History of the Evolution of Life on Earth* (New York: Norton, 2001).

page 155: **Sagan . . . Pollack**: Sagan, *Pale Blue Dot*.

page 158: **"I quite agree how humiliating"**: quoted in Janet Browne, *Charles Darwin*, Vol. 2: *The Power of Place* (New York: Knopf, 2002).

page 158: **Many astronomers and many writers**: For well-written and rather lyrical early takes on this concept, as well as on thermodynamics in general, see Arthur Eddington, *The Nature of the Physical World* (AMS

Press, 1995, reprint of 1927 ed.), especially chapter 4; and James Jeans, *The Universe Around Us* (New York: Macmillan, 1934), especially chapter 7.

page 158: **"The time is the remote future"**: Jack Vance, *Tales of the Dying Earth* (New York: Orb, 2000).

Feeding Time at Loch Ness

page 159: "**Twilight was already**": Murdoch, *The Sacred and Profane Love Machine.*

The Evening and the Morning

page 162: **the evening star**: William Sheehan and John Westfall, *The Transits of Venus* (Amherst, NY: Prometheus, 2004).

Two Lights

page 167: "**Can you feel**": Rainer Maria Rilke, from "The Schmargendorf Diary," in an entry dated 23 July 1898, in *Diaries of a Young Poet*, translated by Edward Snow and Michael Winkler (New York: Norton, 1997). The line break occurs where I divide the sentence with "murmurs Rilke."

Nessie and the Bat

page 167: **Brownell ... "shy and crepuscular"**: Edith Wharton, *A Backward Glance* (New York: Scribner's, 1934).

page 168: **bats**: Clive Roots, *Animals of the Dark* (New York: Praeger, 1974); Merlin Tuttle, *America's Neighborhood Bats* (Austin: University of Texas Press, rev. ed., 1997); Willy Ley, *Dawn of Zoology* (Englewood Cliffs, NJ: Prentice-Hall, 1968).

page 170: **Loch Ness Monster**: Henry H. Bauer, *The Enigma of Loch Ness: Making Sense of a Mystery* (Carbondale: University of Illinois Press, 1986); Steuart Campbell, *The Loch Ness Monster: The Evidence* (Aberdeen, UK: Aberdeen University Press, 1991).

page 171: "**It had grown darker**": Charles Dickens, *Our Mutual Friend*, 1864–1865.

The Twilight Zone

page 172: **Rod Serling**: Gordon F. Sander, *Serling: The Rise and Twilight of Television's Last Angry Man* (New York: Dutton, 1992); Mark Scott

Zicree, *The* Twilight Zone *Companion* (Los Angeles: Silman-James, 1992);
Lynn Neary, NPR report, broadcast 16 September 2002.

page 174 **"There is a long twilight"**: Melville Davisson Post, "A Twilight
Adventure," in *Uncle Abner: Master of Mystery* (New York: Appleton,
1918).

Three Kinds of Twilight

page 174: **Kinds of Twilight**: U.S. Naval Observatory, "Rise, Set,
and Twilight Definitions," online at http://aa.usno.navy.mil/faq/docs/
RST_defs.html.

Memory Has Left the Sky

page 176: **"Memory has left"**: Yvonne Vera, *Without a Name* and *Under
the Tongue* in one volume (New York: Farrar, Straus and Giroux, 2002).

Afraid of the Dark

page 177: **Dewdney . . . size of nighttime**: Christopher Dewdney,
Acquainted with the Night: Excursions through the World after Dark (New
York: Bloomsbury, 2004).

page 178: **"Then comes that mysterious time"**: Nikolai Gogol, "Nevsky
Prospect," in *The Collected Tales of Nikolai Gogol*, trans. and ed. Richard
Pevear and Larissa Volokhonsky (New York: Pantheon, 1998).

page 178: **distribution of public lighting:** Wolfgang Schivelbusch, *Dis-
enchanted Night: The Industrialization of Light in the Nineteenth Century*,
translated by Angela Davies (Berkeley: University of California Press,
1988).

page 179: **As early as the 1500s:** A. Roger Ekrich, *At Day's Close: Night
in Times Past* (New York: Norton, 2005).

The Changing of the Guard

page 186: *cryptochrome*: Randy J. Thresher et al., "Rose of Mouse Cryptochrome Blue-Light Photoreceptor in Circadian Photoresponses," *Science*
282: 5393 (20 November 1992); Russell N. Van Gelder, "Non-visual Ocu-
lar Photoreception," *Ophthalmic Genetics* 22:4 (December 2001).

Lag

page 187: **Jet lag results**: Anna Kuchment et al., "A Good Flight's
Sleep," *Newsweek*, 21 March 2005.

page 188: **pineal gland**: Julia Grosse and Michael H. Hastings, "A Role for the Circadian Clock of the Suprachiasmatic Nuclei in the Interpretation of Serial Melatonin Signals in the Syrian Hamster," *Journal of Biological Rhythms* 11:4 (1996).

page 189: **produces melatonin during the night**: William C. Dement and Christopher Vaughan, *The Promise of Sleep* (New York: Random House, 1999).

page 192: **"Time makes everything old"**: Zora Neale Hurston, *Their Eyes Were Watching God* (New York: HarperCollins, 1990).

Darwin's Busy Plants

page 195: **the book that interests us**: Charles Darwin, assisted by Francis Darwin, *The Power of Movement in Plants* (London: John Murray, 1880); Charles Darwin, *The Collected Papers of Charles Darwin*, ed. Paul H. Barrett (Chicago: University of Chicago Press, 1980); Frederick Burkhardt and Sydney Smith, general eds., *The Correspondence of Charles Darwin* (Cambridge, UK: Cambridge University Press, var. dates); Ritchie R. Ward, *The Living Clocks* (New York: Knopf, 1971); Browne, *Charles Darwin*, Vol. 2.

page 196: **"If asked to choose"**: Barbara Gillespie Pickard, introduction to *The Power of Movement in Plants* (New York: Da Capo, 1966); Takao Kondo and Masahiro Ishiura, "The Circadian Clocks of Plants and Cyanobacteria," *Trends in Plant Science* 4:5 (May 1999).

page 196: **"Circadian regulation of leaf"**: by Andrea Nardini, Sebastiano Salleo, and Sergio Andri, in *Plant, Cell and Environment* 28:6 (June 2005).

page 200: **"You can think better"**: Mati Unt, *Things in the Night*, translated by Eric Dickens (Normal, IL: Dalkey Archive, 2006).

Pillars of Light

page 200: **on cold winter nights**: Keen, *Skywatch East*.

Pale Fire

page 202: **"The line of the horizon"**: Kenneth Grahame, *The Wind in the Willows*, 1908.

The Parish Lantern

page 204: **"the parish lantern"**: A. Roger Ekirch, *At Day's Close: Night* *in Times Past* (New York: W. W. Norton, 2005); the phrase was once a common term for the moon in parts of Britain.

page 204: **"brave mooneshine"**: Samuel Pepys, *Diary*, 1600s. He uses this term in various entries; most editions update the spelling.

page 204: **Muslims begin Ramadan**: Malek Chebel, *Symbols of Islam*, translated by Cybele Hay (New York: Assouline/Barnes and Noble, 2003).

page 205: **The moon lurks everywhere in Japanese prints**: Gabriele Fahr-Becker, *Japanese Prints* (New York: Taschen, 2000).

page 205: **the work of Yonejiro Yoshitoshi**: John Stevenson, *Yoshitoshi's One Hundred Aspects of the Moon* (San Francisco Graphic Society, 1992); Shinichi Segi, *Yoshitoshi: The Splendid Decadent* (Tokyo: Kodansha, 1985).

A Rock in the Sky

page 206: **"Queen of mirrors"**: John Wain, "Moondust," in *Letters to Five Artists* (London: Macmillan, 1969).

page 206: **Nabokov's poem/novel/mock monograph**: Vladimir Nabokov, *Pale Fire* (New York: Putnam, 1962).

The Archangel's Telescope

page 207: **"apologist of all apologists"**: *The Journals of Ralph Waldo Emerson*, ed. Edward Waldo Emerson and Waldo Emerson Forbes (Boston: Houghton Mifflin, 1909–1914), entry for 6 June 1841.

page 207: **In an essay he made even grander claims**: Emerson, "History," 1841.

page 208: **Galileo**: James Reston, Jr., *Galileo: A Life* (New York: Harper-Collins, 1994); Dava Sobel, *Galileo's Daughter: A Drama of Science, Faith and Love* (London: Fourth Estate, 1999).

page 209: **Johannes Kepler . . . Tycho Brahe**: Ferguson, *Tycho and Kepler;* Gingerich, *The Eye of Heaven*; Rudolf Kippenhahn, *Bound to the Sun: The Story of Planets, Moons, and Comets*, translated by Storm Dunlop (New York: Freeman, 1990).

page 210: **"a spyglass by means of which"** and all other quotations from Galileo: Galileo Galilei, *Discoveries and Opinions of Galileo*, translated with introduction and notes by Stillman Drake (New York: Doubleday/Anchor, 1957).

page 210: **the word *telescope***: Richard Panek, *Seeing and Believing:*

How the Telescope Opened Our Eyes and Minds to the Heavens (New York: Viking, 1998).

page 212: **Keats's complaint**: John Keats, in *Lamia*, 1820: "Do not all charms fly / At the mere touch of cold philosophy? / There was an awful rainbow once in heaven / We know her woof, her texture; she is given / In the dull catalogue of common things. / Philosophy will clip an angel's wings, / Conquer all mysteries by rule and line, / Empty the haunted air, and gnomed mine— / Unweave a rainbow." Keats was referring particularly to Isaac Newton's spectrum experiments.

Sunrise on the Moon

page 216: **A day on the moon**: Chet Raymo, *The Soul of the Night: An Astronomical Pilgrimage* (Englewood Cliffs, NJ: Prentice Hall, 1985).

page 217: **to peek at its shy side**: H. J. P. Arnold, ed., *Man in Space: An Illustrated History of Space Flight* (New York: Smithmark, 1993).

The Cup of the Immortal Fluid

page 218: **analog of Amrita**: Heinrich Zimmer, *Myths and Symbols in Indian Art and Civilization* (Princeton, NJ: Princeton University Press, 1972). Not surprisingly, *Amrita* and *ambrosia* are etymological cousins.

Eclipse

page 219: **"And as for me"**: Arno Schmidt, *Scenes from the Life of a Faun*, translated by John E. Woods (New York: Marion Boyars, 1983).

page 219: **eclipse**: Ian Ridpath, ed., *The Illustrated Encyclopedia of the Universe* (New York: Watson-Guptill, 2001); Chet Raymo, *An Intimate Look at the Night Sky* (New York: Walker, 2001).

page 224: **"Aaron saw animal shapes"**: Jerry Dennis, *It's Raining Frogs and Fishes: Four Seasons of Natural Phenomena and Oddities of the Sky* (New York: HarperCollins, 1992).

page 225: **Japanese poet Sodo**: Asatari Miyamori, ed., *An Anthology of Haiku Ancient and Modern* (Tokyo: Maruzen, 1932).

Infinite

page 227: **"Earth cannot"**: Franz Pfeiffer, ed., *The Works of Meister Eckhart* (Whitefish, MT: Kessinger, 1997).

The Lion or the Lobster?

pages 227–228: **Szymborska once remarked**: Wislawa Szymborska, from *View with a Grain of Sand*, translated by Stanislaw Baranczak and Clare Cavanagh (New York: Harvest/HBJ, 1995).

page 228: **"The apportioning of stars"**: Baker, *A Box of Matches*.

page 229: **ethnologist Theodor Koch-Grünberg**: E. H. Gombrich, *Art and Illusion: A Study in the Psychology of Pictorial Representation* (Princeton, NJ: Princeton University Press, 1960); John Bierhorst, *The Mythology of South America* (New York: Morrow, 1988).

page 231: **"The Greeks, who thought"**: Joseph Wood Krutch, *The Twelve Seasons: A Perpetual Calendar for the Country* (New York: Sloane, 1949).

Inconstant Star

page 233: **Polaris**: Ian Ridpath, ed., *The Illustrated Encyclopedia of the Universe* (New York: Watson-Guptill, 2001).

page 234: **literature's most famous invocation**: Isaac Asimov, "Constant as the Northern Star," in *Of Matters Great and Small* (Garden City, NY: Doubleday, 1975).

Mean Time

page 238: **system of radio interferometry**: Ridpath, ed., *The Illustrated Encyclopedia of the Universe*.

The Black of Night

page 242: **Why is the sky dark at night?**: Edward Harrison, *Darkness at Night: A Riddle of the Universe* (Cambridge, MA: Harvard University Press, 1987); John D. Barrow, *Between Inner Space and Outer Space* (Oxford: Oxford University Press, 1999); Raymo, *The Soul of the Night*.

page 243: **"a succession of fainting fits"**: Flann O'Brien, *The Third Policeman* (Normal, IL: Dalkey Archive, 1999).

Swimming Allegories

page 253: **Passamaquoddy song**: T. C. McLuhan, ed., *The Way of the Earth: Encounters with Nature in Ancient and Contemporary Thought* (New York: Touchstone, 1994).

page 253: **John Updike's novel**: *Rabbit Is Rich* (New York: Knopf, 1981).

page 253: **An Ewe song**: Translated by Ulli Beier, in Alan Lomax and Raoul Abdul, eds., *3000 Years of Black Poetry* (New York: Dodd, Mead, 1970).

page 253: **Baker's novel**: Baker, *A Box of Matches*.

page 253: **Vladimir Nabokov**: *Despair* (New York: Capricorn, 1965); *The Real Life of Sebastian Knight* (Norfolk, CT: New Directions, 1941); *Pnin* (Garden City, NY: Doubleday, 1957).

page 254: **one of his short stories**: Vladimir Nabokov, "Wingstroke," translated by Dmitri Nabokov, in *The Stories of Vladimir Nabokov* (New York: Knopf, 1995).

INDEX

Abenaki, 112
absorption, 13, 63–64
acid rain (acid precipitation), 134,
    135–36
Adam Haberberg (Reza), 20
Adams, Ansel, 78
aerial perspective, 63–65
Aesop, 170
Africa, 40, 46, 115, 170, 253
After Dark, My Sweet, 22
afternoon, 106, 107, 116–17, 141
    summer, 136, 137–39
    temperature in, 116, 117–18
Agni, 19
AIDS, 53
air, *see* atmosphere
air pollution, 123–25, 129,
    130–36
    acid rain, 134, 135–36
    ozone, 11, 123–30, 134,
        135, 137
    particulate matter, 123, 134–35,
        137, 227–28
    smog, 13, 124, 131–32
Alice's Adventures in Wonderland
    (Carroll), 39–40, 133
Allen, Woody, 27, 143
All Saints' Day, 107, 160
Almagest (Ptolemy), 59, 228–29
altocumulus clouds, 97
altostratus clouds, 97, 147
"Amanecer" (Borges), 7–8
amber, 260
amphibians, 170
Amrita, 218
Anaximenes of Miletus, 80
animals:
    daily cycles and, 184–85
    twilight and, 164, 168–70

aphelion, 33
apogee, 33
Apollo, 43–49, 51, 54, 71, 206
    epithets for, 43
    Phaethon and, 46–49, 66–67,
        150–52, 181–83
Apollonian, 44
Apollo space program, 31, 44
archaeologists, 240
Arekuna, 229–30
Aristarchus of Samos, 61
Aristotle, 20, 21, 53
art:
    Christian, 14, 145–46, 169
    light and shadow in, 65, 72,
        74–78
    sun as represented in, 3,
        18–20
Artemis, 45, 206
Art of Painting (Pacheco), 14
Asimov, Isaac, 235
Astronomical Unit (AU), 33–35
astronomy, xviii, 88–89, 209, 211,
    247
    constellations in, 229, 231
    Earth's tilt and, 238
atmosphere:
    light rays and, 10–15, 63–66
    organisms in, 64
    ozonosphere in, 11, 119,
        127–30
    pollution in, *see* air pollution
    thinness of, 11
atmospheric perspective, 63–65
Augustus, 232
Aurora, 26, 49
aurora borealis and aurora australis,
    157
Aztec mythology, 115, 156